St. Louis Community College

Library

5801 Wilson Avenue
St. Louis, Missouri 63110

DISCRIMINATION, PERSONALITY, AND ACHIEVEMENT
A Survey of Northern Blacks

DISCRIMINATION, PERSONALITY, AND ACHIEVEMENT

A Survey of Northern Blacks

**Robert L. Crain
and Carol Sachs Weisman**

*Department of Social Relations
The Johns Hopkins University
Baltimore, Maryland*

SEMINAR PRESS New York and London 1972

SEMINAR PRESS, INC.
111 Fifth Avenue, New York, New York 10003

United Kingdom Edition published by
SEMINAR PRESS LIMITED
24/28 Oval Road, London NW1

LIBRARY OF CONGRESS CATALOG CARD NUMBER: 72-10104

PRINTED IN THE UNITED STATES OF AMERICA

CONTENTS

4. Aggression

5. Internal Control

6. Self-Esteem

7. Happiness Is . . .

8. Personality and Achievement

9. The Broken Home

10.^{X} Discrimination, Personality, and Achievement

11. School Integration: An Evaluation

12. The Policy Issues: An Essay

Preface

The period from 1954 to 1964 was a peculiarly satisfying time in American race relations—when people knew not only where other people stood, but even what they themselves believed. It is rare for an issue to be so clearly drawn. The decade since then has been a bitterly dissatisfying time. The issue of race became more muddy as the media became filled with angry, pessimistic reinterpretations of events, each in turn subject to immediate and vicious counterattack. Now it seems none of us knows what to believe or whom to trust.

At one level, the argument has been about the merits of assimilation *versus* cultural pluralism. But more important, it has become permissable for whites to talk about black inferiority, and blacks to express hatred toward whites. The result is a chaotic set of cross-cutting arguments about questions whose answers were taken for granted ten years ago:

> *Why* do blacks have to be integrated?
> Are blacks *really* inferior?
> Why *do* blacks commit so many crimes?
> White racist?! Who, *me?*

Perhaps it is better to have stopped "sweeping" these questions under the rug. The next task is to try to answer them since they are not going to disappear.

This book is an attempt to produce a scientific analysis which will clarify—though hardly settle—some of these questions. The study began as an analysis of the long-term effects of desegregated schooling on blacks, as requested by the U.S. Civil Rights Commission. What we found was confusing. The evidence indicates that desegregated schooling did increase the chances for blacks to go to

ix

college, but it did little to increase performance on academic achievement tests. More surprising, school integration had a striking long-term effect on some aspects of black personality. For example, blacks who attended northern segregated schools were much less happy than those who attended integrated schools.

For an explanation, we turned to theories of black personality developed in the late 1930's by several anthropologists, psychologists, and psychiatrists—John Dollard, Allison Davis, Hortense Powdermaker, and, in particular, Abram Kardiner. Briefly, their general thesis was that widespread discrimination frustrates blacks, damaging their self-esteem and creating a need for them to express aggression in reply. With damaged self-esteem and uncontrollable aggression, blacks would be hard put to succeed in life even if given the opportunity. This theory helps explain the present disabilities of blacks and suggests why integration is beneficial. According to this theory, segregation, as a symbol of lack of opportunity, is as harmful as discrimination, the actual restriction of opportunity.

A plausible thesis, but is it correct? To test it, we used the rigorous techniques of survey research, interviewing large numbers of blacks (and for purposes of comparison, whites) with a standard questionnaire, and analyzing the results with a statistical analysis. Moreover, we applied these techniques to problems usually within the province of clinical psychology, which rarely deals with "normal" populations.

Chapter 1 describes the methods of research in more detail. Chapter 2 introduces the reader to the survey results. It describes statistically many of the problems of blacks and points out that the massive black migration from the South to the North seems to have been harmful in as many ways as it was helpful. Chapter 3 describes Abram Kardiner's theory in more detail, and examines the writings of Martin Luther King, Jr., to show that a similar theory is implicit in much of his thought.

Next, we studied the questionnaire responses to determine if we could find in them indications of certain kinds of personality difficulties and evidence that these personality characteristics do affect a person's chances of success. First we used the questionnaires to identify the characteristics of those people who, from their responses, seem to have trouble dealing with aggression: either they express aggression too openly, or, being unable to express anger, are too inhibited. A second personality trait examined in Chapter 5 is the lack of a sense of internal control of environment—the belief that one cannot control one's future; that life is dominated by blind chance. In Chapter 6 we examine the question of black self-esteem and in Chapter 7 the responses of blacks about their sense of being happy.* The findings suggest that blacks do have more difficulty than whites with aggression, are more fatalistic about the future, and are much less happy.

*Carol Weisman was responsible for the analysis and writing of Chapters 6 and 7.

However, blacks born in the North do *not* have lower self-esteem than whites. In Chapter 8, we begin to collate these four characteristics and find that they describe two kinds of responses to life: the aggressive self-confident person who is overly fatalistic and unhappy and the person who is not fatalistic, not unhappy, but is inhibited and low in self-esteem. We then show that these personality differences do seem to make a difference in one's chances for economic success.

There are two factors in the black experience which might explain why blacks differ from whites. First is the high rate of marital separation and divorce among the parents of our respondents. The other is, of course, racial discrimination and segregation. We find some rather convincing evidence that marital instability (Chapter 9) and racial discrimination (Chapter 10) are harmful. Men who come from broken homes tend to have problems controlling aggression and lead rather unstable lives with quite low lifetime earnings.* Men who grew up in the South are more fatalistic and have very low self-esteem. Black men who grew up in the North but attended segregated schools have high self-esteem, but are also fatalistic and quite unhappy as adults. Other factors related to the respondent's racial experience also seem to make a difference.

The overall pattern is somewhat different from that predicted by Kardiner. Southern discrimination does lower self-esteem; but blacks who were born in the northern ghettos react to segregation not by internalizing their feelings and thinking of themselves as worthless, but by externalizing them, going about in an unhappy rage.

In Chapter 11, we analyze segregation in schools, and in the concluding Chapter 12, the issues on national policy raised by this analysis.†

These chapters are hardly the last word on the condition of northern blacks. For every question tentatively answered here, there are a dozen others to which we do not know the answers. And no doubt there are some incorrect answers here as well. Facts may speak for themselves, but they don't say very much; and if we want to know, for example, why black alumni of integrated schools earn more money than those from segregated schools, we must settle for an argument made by stringing together some facts, some assumptions, and some hopefully persuasive discussion. And since no one can make such an argument without his own prejudices crashing through, perhaps we should state our prejudices. We are "nonviolent, integrationist liberals." Long before this study was conceived, we were convinced that racial equality would come through some sort of racial integration. We were also convinced that the importance of racial prejudice and

---

*The analysis of the effect of family structure on women (including the concept of a culturally transmitted trait of "female-dominance") was being done by Carol Weisman at the time this book was published.

†Chapter 12 is solely the work of Robert L. Crain.

discrimination had been ignored by too many writers concerned with the plight of blacks. Nothing in these data has made us question our convictions, and we think our argument is scientifically sound; but the reader will perhaps have an easier time trying to criticize our interpretations if we state our prejudices openly at the outset.

Robert L. Crain
Carol Sachs Weisman

Acknowledgments

It sometimes seems that the concept of authorship is incompatible with the realities of modern social research. Although this manuscript has only two authors its research is the collective work of dozens of people.

Six graduate students at Johns Hopkins and the University of Chicago contributed to the project. At Johns Hopkins, Martha Leatherman analyzed much of the integration data, Michael Ornstein wrote three papers from the data that are incorporated in the analysis of marriage, and Sue Bobrow analyzed the effects of being raised in a broken home. David Klassen, the first student on the project at Chicago, developed the codebook, supervised some of the task of computerizing the study, and did some analysis. Judith Favia, a graduate student at Chicago, was organizer of the study in its first year and developed a routine which kept the study from sinking under the weight of a quarter-ton of computer output.

The study owes its greatest debt to Starling Goodyear Pullum, then a graduate student at Chicago. In the course of a year's conversation, her insights into the social psychology of our respondents shaped the final outline of the analysis.

It is interesting, even though often futile, to consider the origins of the ideas that go into a study. The most important influence in this case is probably that of Fred L. Strodtbeck whose writings and teachings did a great deal to set the frame of reference for this work.

The study was initiated at the National Opinion Research Center by Peter H. Rossi, Paul M. Siegel, and myself. It was Rossi's interest in race relations research which prompted the U.S. Civil Rights Commission to approach NORC regarding this study, which at that time they planned to include in their report *Racial Isolation in the Public Schools*. Paul Siegel and I took over at that point and prepared the questionnaire. It was originally intended that Siegel coauthor the report, but unfortunately the press of other research made this impossible.

The questionnaire was prepared and administered under the direction of NORC's survey research service. The final draft of the questionnaire was prepared by two highly skilled questionnaire draughtsman, Eve Weinberg and Carol Bowman Stocking.

The staff of black interviewers was hired by Mrs. Weinberg and Fansayde Calloway. The field supervision was by Celia Homans; her illness in the later stages of the field work was a serious loss. The coding was supervised by Francis Harris. The computer work at NORC was directed by T. Russell Shields. Nancy Morrison managed the computer processing, using a program developed by Jim Jasper.

At Johns Hopkins, the computer work was handled by Narindar Kelly, using the installation operated by Nancy Karweit, assisted by Hao-Mei Kuo. Nancy's programming and administrative skills were invaluable.

Dolores Sullivan and Barbara Curtin typed the manuscript; thanks are also due the NORC stenographic pool, supervised by Toshiko Takahashi. Patrice Kraus was a valuable research assistant and editor.

The study was funded initially by the U.S. Civil Rights Commission; William L. Taylor and David Cohen were administrators of the project and Thomas Pettigrew the prinicpal consultant. The study went "over the budget" and NORC graciously agreed to fund the study out of its working capital. Later, the Ford Foundation provided the funds for the preparation of this monograph. The Ford Foundation was very generous in permitting numerous time extensions and administrative changes. The Johns Hopkins University provided salaries for faculty and fellowships for students while the book was being written. Carol Weisman was supported by a National Institute of Mental Health fellowship, number 1F01 MH 47919-01.

But our greatest debt must be to the interviewers and to the 1651 respondents who so generously surrendered their time and privacy to provide the data.

1
The Research Design

In the South we were always insulted and my parents would always tell us boys to take it because we couldn't win and if we tried to do anything about it we would probably be killed. My father would tell us that the white people had a meanness in them God Almighty himself couldn't get out of them and just go and take it and when we got older to leave the South, which we all did.[1]

In Holmes County, Mississippi, it is said that after high school commencement, graduates marched off the stage to the train station. The train is the Illinois Central, and it runs to Chicago. We can experience a vicarious thrill as we imagine those 18-year-olds climbing down from the coaches at Chicago's 12th Street station, realizing that they never again had to see a cotton field or a sign that says "colored" on a washroom door. Like the Puritans, the Irish, the Italians, and the Jews before them, they were discovering America.

For most of them, the dream did not come true. What many of them did find in the ghettos of the cities was a life of second-rate jobs, second-rate apartments, bad marriages, and illegitimate children. Why? There are short answers, none of them satisfactory. Discrimination and white racism are part of the answer, but it is not a complete lie to say that many employers today are begging for showcase blacks and there are too few to be found. Racial inferiority is the oldest answer, having been applied to the Irish and Poles as well, but it is not scientifically sound. The truth, as usual, is much more complicated.

[1] From the questionnaire of a 40-year-old man born in Louisiana, who came North when he was 17 and now lives in Los Angeles.

1

The Study

In the Spring of 1966, the U.S. Civil Rights Commission asked the National Opinion Research Center (NORC) to conduct a survey of northern blacks to determine the effects, if any, of attending integrated versus segregated schools. The result was an extensive survey of 1651 black men and women, aged 21 to 45, living in the metropolitan areas of the North. This is how the study began. We were also interested in broader questions, and the study became a survey of the state of northern blacks.

The sample for this survey included 297 randomly selected "city blocks" in 25 different metropolitan areas in the North. The interviewers, all of whom were black, were assigned to selected blocks and told to systematically locate six black men and women in each block who were willing to be interviewed. This type of sampling procedure is known as "cluster" sampling, since it uses randomly selected clusters of city blocks within census tracts also chosen by a random procedure.[2] The sample was limited to northern metropolitan areas since so few southern adults have attended integrated schools.

The sample is "weighted," because it contains more high-income respondents and more respondents from small metropolitan areas than it should. The low-income and big-city respondents were then weighted—counted more than once— to make the statistics representative of the total population. The weighting, which was designed to make our analysis of high income and small-city blacks more accurate, has such a slight effect on our statistics that no explicit attention will be given to it in the analysis. The number on which our percentages are based is 2½ times the actual number of respondents interviewed: the 1651 people in our sample appear as if they were 4153. The reader who is interested in statistical significance should routinely divide the table case bases by 2½ to get an approximation of the true number.

In a study that purports to analyze northern American blacks, some comparative statements must be made to point out the differences between the black population and the white population. For this reason many of the questions included in the questionnaire were also included in another NORC national survey, which included 1326 white respondents. This national sample of whites will serve as a basis for comparison throughout this book.

A study of the effects of school integration involves many facets of the lives of black men and women. Their school experience might affect their chances of going to college, their abilities to get jobs, their feelings about white people, their attitudes toward integration, their self-image and motivation, and in turn, all of these might affect other parts of their lives, such as their marriages. By the time

---

[2] Appendix 1 contains a more complete description of the sample and an analysis of the sample bias.

we finished adding questions about the other areas in which we were interested, the final version of the interview took an average of 2 hours and contained nearly 500 questions and subquestions. Many questions were "precoded"—could be answered simply by "yes" or "no" or by the use of other categories—but sometimes the complete answer was recorded verbatim. The most important instruction given to the interviewer was to read the question exactly as written and to avoid any comments which would suggest how the respondent should answer.

What Can We Learn from a Survey

It is tempting to think that we did what we seemed to be doing—finding out the "true" answers to a set of questions. Certainly we are dealing with truth, but a specific kind of truth. For example, we asked the respondents if they had ever been arrested and 22% said "yes." The seasoned survey researcher will read this number and conclude that probably somewhere between one-fifth and one-half of our respondents have indeed been arrested, depending on how we define arrest, how our respondents interpret the word, how many people are mistaken, and how many lied. The important conclusion we can draw is that the 22% who said "yes" are more likely to have been in trouble with the police than the 78% who said "no." This simple-minded conclusion is nevertheless useful. For example, men are much more likely than women to say "yes," and blacks much more likely than whites. In this case it seems safe to conclude that black men are indeed arrested more frequently. But in some cases, where our knowledge is limited, even such a simple-minded sort of conclusion is not possible. We know, for instance, that young men are more likely than older ones to say "yes." Does this mean that the younger generation is having more trouble with the police than the 40-year-olds? It probably does, but it may just mean that the younger men are honest, less forgetful, or more aggressive in their relationship with the interviewer.

Drawing conclusions from the data can therefore be a tricky problem, particularly when the data refer to attitudes or to psychological attributes. When we ask a respondent how he "feels" about something, we can interpret his responses as indicating a psychological state or as showing a predisposition toward a certain kind of behavior, but we cannot interpret the response literally. For example, one-third of our sample said "yes" when asked if they sometimes wanted to "get even" with whites. We cannot conclude that one-third will attack whites, or even that one-third wants to; only that one-third said "yes," and from that point of beginning, it is the task of the analyst to determine how that third differ from the two-thirds who said "no."

The right way to think about the survey is as a set of micro-experiments: the subject is presented with a stimulus, in this case an interviewer reading a question, and he makes a verbal response. It is our job to decide the meaning of a particular response being given more frequently than another. Virtually the only way to interpret these responses is by comparing the responses of two different groups (male and female, old and young, black and white). Thus the task of the analyst and the reader is to ask, "Can we decide what it means that one type of person is more likely than another to give a certain response to a particular question?"

Survey Methods and Personality Theory

The survey method using large national samples is the province of sociology more than psychology. There have not been very many efforts to use large-scale surveys to measure personality traits. The reason is obvious. If a trained psychoanalyst must spend years with a single patient in order to imperfectly understand one personality, any effort to understand anything about personality from a 2-hour interview might seem silly. Furthermore these 2 hours were not a clinical diagnostic interview. It was a formal written interview in order to guarantee strict comparability between subjects. Psychologists do this sort of thing, primarily with college students as subjects, but even there they use lengthy batteries of questions, using scales of 20–40 questions, for example. A survey as expensive as this one cannot afford such careful measurement of one or two concepts; it must "spread itself thin." Thus both clinical and experimental psychologists will find this work unsatisfying. But in fact, it seems to us that despite the errors of measurement, the only way to answer questions about the impact of social change on the human personality is with this kind of methodology. It is probably the only way to move psychology out of the laboratory, to subjects more real than college sophomores, at reasonable cost.

There is no way to develop a perfect scientific proof that, for example, attending integrated schools increases a black's sense of control over his environment. We can show that graduates of integrated schools do in fact score higher on a measure called "internal control" and that this is not because alumni of integrated schools come from middle-class homes. We can show that this measure does seem to reflect a sense of control over one's environment. We can argue that the hypothesis that integration *causes* a higher sense of control is consistent with some psychological writings as well as with some of our other findings. This argument satisfies us, and it should convince the reader. But social research has no way of making a theory into an incontrovertible fact.

The reader may wonder whether a survey done in 1966 has any relevance when it is read years later. We are not so concerned with this. This book is primarily about social relations, economic conditions, and especially personality, and as most of us have found out to our sorrow, these things change very slowly. Indeed, when we look at personality, our theoretical orientation will be taken from a study of New York blacks done over 30 years ago.

Political slogans may come and go, and fads, customs and styles of behavior may change rapidly; fundamental changes in the social order are much slower.

Two final points should be noted. First, it goes without saying that since the analysis involves contrasting blacks and whites, we will not pay much attention to the fact that they are more similar than different. Certainly a black school teacher in Los Angeles is more like her white colleague than like a southern black farmer, but since we are studying blacks as a group, we will pay more attention to the differences between whites and blacks than to the similarities.

Second, the survey methodology means that we are concerned with average responses of groups, and no individual is ever average. The reader is urged not to look for his acquaintances (or himself) in the tables presented, but rather to think in terms of groups of people as they represent trends or types. Note, in addition, that it is not illogical or immoral to look at averages. It is not racism to determine that blacks as a group are more likely than whites to say they have been arrested because it is not legitimate to conclude from this fact that whites are racially superior to blacks. What we can conclude from this fact, and others like it, is that blacks and whites are in different social and psychological niches in society. And then we may begin to ask why.

2
The Failure of the Dream

But suppose we should rise up to-morrow and emancipate, who would educate these millions, and teach them how to use their freedom? They never would rise to do much among us. The fact is, we Southerners are too lazy and unpractical, ourselves, ever to give them much of an idea of that industry and energy which is necessary to form them into men. They will have to go north, where labor is the fashion,—the universal custom; and tell me, now, is there enough Christian philanthropy among your Northern States, to bear with the process of their education and elevation? You send thousands of dollars to foreign missions; but could you endure to have the heathen sent into your towns and villages, and give your time, and thoughts, and money, to raise them to the Christian standard? That's what I want to know. If we emancipate, are you willing to educate? How many families, in your town, would take in a Negro man and woman, teach them, bear with them, and seek to make them Christians? How many merchants would take Adolf, if I wanted to make him a clerk; or mechanics, if I wanted him taught a trade? If I wanted to put Jane and Rosa to a school, how many schools are there in the Northern States that would board them? And yet they are as white as many a woman, north or south. You see, cousin, I want justice done us. We are in a bad position. We are the more *obvious* oppressors of the negro; but the unchristian prejudice of the north is an oppressor almost equally severe.[1]

St. Clare to Miss Ophelia, *Uncle Tom's Cabin*

We don't think we are being harsh or extreme when we say that the North has failed. The dream represented by the North has become a nightmare. Many writers have, in trying to explain this, argued that we must give black migrants

[1] Harriet B. Stowe (1851), *Uncle Tom's Cabin or Life Among the Lowly*, Vol. 2, p. 77, Riverside Press, Cambridge, Mass.

from the South time to adjust to northern urban life, to develop nonagricultural skills, to send their children to urban schools, and to learn how to live in, and take advantage of, the city. Unfortunately, the data from our survey suggest that life in the North is so disruptive that migrants from the South are actually better off than those who were born in the North.

Income

While not in any way surprising, the most striking message that comes from the survey is that many blacks are poor. The median income for northern black families was $5700 in 1966, only 70% of the median income for northern whites.[2] Poverty does not necessarily mean unemployment and starvation—few of our women were on welfare, and few of our men were unemployed. It does mean economic insecurity: not being able to own a house, being in debt, and having at least an occasional catastrophe. Thirty-eight percent of the sample have neither a checking account nor a savings account. For most of these, this means they have no savings at all; the smallest emergency requires borrowing. Most of our respondents are in the red: 66% have debts which they could not now pay off. No doubt most of these will be able to make their payments, but again, there is no margin for emergencies. A sizable number of people have suffered financial catastrophe at some time. Twenty-five percent have had their utilities turned off, compared to only 8% of whites of the same age range in northern cities.[3] Ten percent have been evicted, compared to 3% of whites.

We need not assume that this is the result of bad money management. Mostly it is a direct consequence of low income. In 1950, the national ratio of nonwhite to white median family income was .54; in 1963 it was .53.[4] Large numbers of blacks moved out of the South in those 13 years; if it had not been for this migration into high-paying northern areas, blacks would have actually lost ground relative to whites. It was not until the late 1960's that blacks showed any income gain relative to whites. Much of this income discrepancy seems due to job discrimination. For example, the income gap is sharpest among the college educated; in 1959, the mean yearly income for black college graduates was $5671, only 55% of the white mean. Black high school graduates, with a mean of $4021, earned 64% of the white mean.[5] If the qualifications of black employees were a major factor, we would not expect black college graduates to be at the greatest relative disadvantage.

[2] Figures from Herman Miller and Dorothy K. Newman, (October 1967), *Social and Economic Conditions of Negroes in the United States,* p. 16, joint publication of the Bureau of the Census, U.S. Gov't. Printing Office, Washington, D. C.

[3] Data from the comparison survey of whites.

[4] *Ibid,* p. 15. A sharp improvement has occurred since 1963; the ratio was .64 in 1970.

[5] Paul M. Siegel, (1965), "The Cost of Being Negro," *Sociological Inquiry* **35**, 41-57.

White-black hiring differences are sometimes brought to public view by suits against large manufacturers or construction unions. Much less visible is the natural (and economically sound) tendency of firms to recruit new employees from friends and relatives of present employees; if the present staff is white, these friends and relatives will also be white. Also important in job discrimination is the fact that small employers in white neighborhoods hire many of their staff from "walk-ins" who live in the neighborhood. Thus, even if actual discrimination in all hiring decisions ended tomorrow, these other mechanisms could maintain inequality in job opportunities for a long period of time.

Discrimination is only one of the factors in black poverty. The unstable family is another. The median family income for married black couples in 1965 was $7470 per year, considerably higher than the median income for all black families. This high income is partly due to the presence of working wives, but it is also because married men have higher individual income than divorced men. The fact that divorced men earn less than married men could mean three things: it could mean that the breakup of a marriage causes trauma, dislocation, and freedom from financial need so that men who separate from their families are less able or willing to hold jobs or work overtime. It might also be that the characteristics which make it difficult for a man to earn a satisfactory income are also characteristics which make it difficult for him to stay married, or simply that low income is damaging to the marital relationship.

The personal incomes of working women are much lower than those of men. And of course formerly married women have very low family incomes, since they are based almost entirely on either their own salary or public assistance. The high rate of marital breakup, the low incomes of women, and the simple fact that one cannot earn as much as two, combine to produce a sizable amount of black poverty. Six-hundred thirteen cases (15% of the sample) are formerly married women. Most have children and one-half have family incomes below $4080.

The Melting-Pot Urban Immigrant Hypothesis

We must consider the possibility that these figures represent "rapid" improvement over the last generation—as rapid as can reasonably be expected. Put more formally, any social system has limited opportunities for social mobility; indeed the evidence seems to indicate that there is about the same amount of economic mobility in most industrial societies.[6] This argument then says that given their starting point one generation ago, blacks have done about as well or as poorly as any other disadvantaged group in an industrial society. Is this a plausible hypothesis?

[6] Seymour Martin Lipset and Reinhart Bendix, (1959), *Social Mobility in Industrial Society,* Univ. of California Press, Berkeley.

The classic assimilation theory as applied to blacks parallels their migration north with the arrival of earlier immigrants. They are peasants with a somewhat foreign culture, little education, and little experience with industrialization or the urban environment. The European immigrants arrived in a port-of-entry neighborhood of the city, took the worst jobs at the lowest pay, and under the initial shock of cultural change, families became more disorganized, children became delinquents, and the "good people" of the city wondered about the filthy habits of the newcomers. However, each immigrant group managed to gradually adjust and pull itself up the ladder. The argument states that blacks in the North are the new immigrant group, and that the same processes that operated in earlier groups are at work to bring about their assimilation.

The best description of the classic assimilation theory is Louis Wirth's study of the Jewish ghetto in Chicago from the 1880's to the 1920's.[7] He views the process of assimilation as an inevitable one that occurs to all immigrant groups. According to Wirth, the first generation (the immigrants themselves) settle in lower class areas in their ports-of-entry because rent is cheap and they can live close to their own kind. Remaining isolated in these ghettos, they are unaware of their deprivation relative to other American communities. The second generation (their children) then grows up and begins making contacts outside of the ghetto. The second generation becomes partially assimilated, and somewhat marginal in its cultural identity, since it is caught between the old traditions and the lure of the outside community. This "anomic" condition of being caught between two cultures causes this generation to undergo considerable strain. By the time the third generation comes along, however, the assimilation process is complete; the third generation does not grow up in the ghetto.

What Wirth is referring to here as "assimilation" is less the abandonment of an ethnic identity than it is the adoption of a middle-class (or upper working-class) American lifestyle. Virtual assimilation has not occurred to immigrant groups since they have not lost their ethnic identities in becoming American. When Wirth's second and third generation Jews moved out of the Chicago ghetto they escaped their impoverished surroundings, but they remained a distinct group. The ethnic identity was maintained while the socioeconomic status of the group changed. Assimilation is really a euphemism for upward social mobility.

No one has argued that the three-generation assimilation process should apply precisely to blacks. Unlike the late 19th century immigrants, they come to the North as English-speaking American citizens, and come from a region whose culture is not greatly different. It seems likely that this would speed up the adjustment process. In addition, the presence of mass media—first radio, then television—should help break down the cultural parochialism of the first-generation migrants. But if blacks are like other ethnic groups, we should find that migration

[7]Louis Wirth (1928), *The Ghetto*, Univ. of Chicago Press, Chicago.

produces a trauma, followed by a period of maladjustment, fading into the beginnings of successful adaptation to life in the northern city, although perhaps not in exactly three generations.

The assimilation theory as applied to blacks is an optimistic one, for it argues that much of the troubles we presently observe are temporary: the result of the presence of many second-generation immigrants, who have been forced to rebel against their parents' customs, which were learned in the old country of the South, and are irrelevant in the new world of the North. The next generation, according to the theory, should make the beginnings of a successful adjustment.

There is a certain amount of plausibility here. Our respondents are primarily second-generation: only 45% were born in the rural south, and almost half the respondents grew up in the North. Only 20% give their father's occupation as farming. The typical pattern is parents who came to the North or to a southern city either before our respondents were born or shortly afterwards.

It is also true that the next generation will not experience as much "cultural deprivation" as our respondents have. For example, 36% of our respondents come from homes with eight or more other children, and only 19% have fathers who finished high school. (For whites in this age group the figure is 39%.) In terms of family size and parents' education, the black children being born now should not be greatly different from whites.

But before we accept this untestable prophesy that the troubles blacks are having are temporary, let us take a close look at the data we have available. It seems to us that if the assimilation model were accurate, we should find a significant number of northern-born adults are successful at this time, and in particular, that the youngest group of northern-born respondents (many of whose parents have lived in the North since the beginning of the World War II) are overcoming the shock of immigration.

Migration and Income

If the classical model of immigration is valid, those blacks who have been in the North longest should have learned the secrets of the city, thrown off their southern ways, and made the most progress. In fact, northern-born blacks do *not* have noticeably higher incomes than immigrants from the South (Table 2.1). We have divided immigrants into two groups: the "early" migrants, who came north before they were 10-years-old, and hence received nearly all their schooling in the North, and the "late" migrants who came north after age 10. Almost all of these came north as adults, either just after leaving high school, or, for many men, on completion of military service. Unfortunately, there are not very many early migrants in the sample, so this category is often difficult to interpret due to the possibility of large sampling errors.

TABLE 2.1

Incomes of Southern- and Northern-born Respondents by Sex, Age, and Education[a]

Sex, age, education	Median individual income (dollars)		
	Place of birth, age of migration		
	Born in North	Moved North before age 10	Moved North after age 10
All men	5466 (590)	5564 (232)	5598 (900)
All working women	3842 (293)	3272 (127)	3117 (453)
MEN			
Age 30–45			
High school graduate	6720 (186)	5792 (67)	6859 (171)
Not graduate	5191 (125)	4750 (67)	5526 (471)
Age 21–29			
High school graduate	4975 (208)	5905 (57)	5281 (136)
Not graduate	4450 (71)	4937 (41)	4984 (122)
WORKING WOMEN			
Age 30–45			
High school graduate	4900 (96)	4000 (33)	4100 (136)
Not graduate	3667 (71)	3200 (40)	2801 (193)
Age 21–29			
High school graduate	3767 (84)	4700 (26)	3357 (86)
Not graduate	2125 (42)	2000 (28)	2429 (38)

[a]Numbers in parentheses are figures on which median individual income was based.

For men, both early migrants and late migrants have slightly higher incomes than native Northerners. This is in spite of the fact that northern-born men have much better educational opportunities than the migrants. Sixty-five percent of the Northerners are high school graduates, compared to 55% of the early migrants and 34% of the late migrants. Let us separate the effects of education from the region by adding education as a "control variable" in Table 2.1. (We have also controlled on age, since migrants are older than the native Northerners.) In the table each sex is divided into those over and those under age 30, and those who did and did not graduate from high school, so that there are four comparisons to make for each sex. Older men and women who finished high school have the highest incomes; for example, native northern high school graduate men over age 30 have a median income of $6720.

Men who were late migrants from the South earn $100–$500 more per year than northern-born men in each age and education category. The early migrants from the South earn about as much as northern-born men. Furthermore, it is the younger migrants, especially those who are not high school graduates, who have the greatest economic advantage over native Northerners. This suggests that it is in the unskilled and semiskilled jobs, where young high school dropouts are most likely to be, that Northerners have the greatest economic disadvantage. Women show the opposite pattern: northern-born women have generally higher incomes than those raised in the South. (This is the first of a large number of tables we will present which show opposite patterns for men and women.)

There is a major problem with these data. The Southerners who moved north are not typical Southerners; as many studies have shown, they are the most skilled, and they probably are more ambitious as well. We need data on the Southerners who stayed behind. However, we think the income differences cannot be explained entirely by a selective migration hypothesis. One reason for this is that we cannot explain the superior adjustment of Southerners who came north as children on the grounds that their parents are self-selected and more adaptable than the Southerners who stayed behind, because the vast majority of northern-born respondents are also the children of migrant parents.

The southern migrants are not well educated. As the bases for the percentages in Table 2.1 indicate, only one-half (53%) of the southern-born men who came north after age 10 and were under age 30 in 1965 are high school graduates, and only one quarter (27%) of those over 30 have diplomas. The comparable percentages for northern-born respondents are 75% and 60%. If the southern migrants who came north before age 10 are a superior group or have more ambitious parents, it doesn't appear in their rates of high school graduation. Only 58% of the younger men and 50% of the older men graduated despite the fact that most of their education was in northern schools. Yet this group of migrants have incomes as high as Northerners, and we will see in the next few tables that they seem to be better adjusted to urban life.

Migration and Antisocial Behavior

Many writers have described the ghetto as an armed camp with three locks on every apartment door. Our respondents were asked, "When you go out into the neighborhood at night, are you sometimes afraid of being robbed or attacked?" Sixty percent of the women and 27% of the men living in all-black neighborhoods said "yes." Integrated neighborhoods are apparently safer; in mostly white areas these percentages drop to 42% and 13% respectively. Whites generally feel that they are the principal victims of black crimes, which is, of course, untrue. In our study, 35% of blacks reported being robbed of as much as $20 at some time, compared to only 26% of northern urban whites.[8] If assimilation into

[8] These figures tend to agree with those printed in *Social and Economic Conditions of Negroes, 1970* (1971), U.S. Government Printing Office, Washington, D.C. The figures there show for example, that black women are five times as likely to be victims of violent crimes as are white women. (The data is attributed to the President's Commission on Law Enforcement and the Administration of Justice.) Other racial differences are smaller.

northern ghettos leads to a positive adjustment, we would expect the more assimilated blacks to be in less trouble with the police. Table 2.2 gives the response to
the question, "Have you ever been arrested?" We would expect that the risk of
arrest should be higher in the North; at least it is commonly said that a black is
less likely to be arrested for a crime committed against another black in the
South (and most crimes committed by blacks are committed against other blacks).
Thus the best comparison this time should be between those who were born in
the North and those who moved to the North as children, spending their teenage
years trying to adapt to the urban environment and under the surveillance of
northern police. But whichever category of migrants is used for comparison, the
data indicate that native Northerners are slightly more likely to have been
arrested than are migrants. In 7 of the 8 comparisons, the migrant men have
lower arrest rates than the corresponding group of native Northerners. Late migrants (column 3) who are over 30 and are high school graduates have a higher
rate of reported arrests, 33%, than native Northerners of the same age and education. Northern-born women have higher arrest rates than migrants in 6 of 8 comparisons in Table 2.2.

When we consider that at least a few men must have refused to admit to an
arrest record, these numbers seem very high. We also see that men under 30 report being arrested as often as those over 30. If this is true (i.e., if there is no
difference in the response bias of the two age groups) it suggests that this younger

TABLE 2.2
Arrest Rate, by Age Moved North, Age, Sex, and Education (% Arrested)[a]

Sex, education, age	Age moved to the North		
	Born in North	Before age 10	After age 10
MEN			
High school diploma or more			
Age 30-45	31 (187)	14 (71)	33 (174)
Age 21-29	30 (228)	25 (61)	27 (150)
Without high school diploma			
Age 30-45	64 (143)	49 (71)	41 (523)
Age 21-29	55 (87)	54 (41)	40 (130)
WOMEN			
High school diploma or more			
Age 30-45	5 (179)	0 (85)	4 (314)
Age 21-29	13 (172)	7 (56)	4 (237)
Without high school diploma			
Age 30-45	14 (159)	18 (79)	13 (559)
Age 21-29	8 (158)	10 (98)	4 (143)

[a]Numbers in parentheses are figures on which percentages were based.

generation in another decade will have added enough additional arrests (especially for such adult crimes as drunkenness) to surpass their elders. The arrest rate may be going down slightly over time due primarily to the much higher educational attainment of the younger age group. But the overall picture is not very encouraging.

The reader may wonder whether the failure of arrest rates to drop in the younger generation may not be due to increased law enforcement in the ghetto. But respondents were also asked, "Have you ever been in a fight—not an argument, but a real fight—since you've been an adult?" Over a third of our males said, "yes," and men aged 21-29 reported having been in a fight almost as much as did older men. It seems very likely that by the time the present group of men in their twenties are as old as the older men in our sample, more of them will report fighting. If we ignore reporting bias, this suggests that the extent of interpersonal violence may be increasing. At least it seems safe to say that the rate of violence is not going down very fast. Northern-born men are more likely to report being in a fight than are late migrants from the South with the same amount of education. The data on arrests and fights seem to indicate that being born in the North, instead of being an advantage to being born in the South, increases one's chances of being involved in violence.

Migration and Marital Breakup

Daniel P. Moynihan's thesis suggests one possible explanation of the offsetting and cancellation of the advantages that "second generation" Northerners should have. He argues in *The Negro Family: The Case for National Action*[9] that there is a vicious cycle in the big-city ghettos wherein discrimination and lack of skill lead to low income which in turn leads to marital breakups, thus producing fatherless boys who will grow up as unhappy and as unskilled as their parents. Although we shall postpone an empirical investigation of this argument until later, Table 2.3 supports his report of the high rate of marital instability in the ghetto.

Forty-four percent of the women over age 30 in the sample saw their first marriages dissolve through separation and divorce. Under the age of 30, there are sharp differences between migrants and native Northerners: northern-born women are more likely to have been separated or divorced than were late migrants by a ratio of 3-2; for young men the ratio is 5-2. Over the age of 30, the two groups are closer to each other, but the late migrants have consistently lower separation and divorce rates. Note that in each age group the reported divorce and separation rates for men are lower. Part of this is because husbands

[9] Daniel P. Moynihan (1965), *The Negro Family: The Case for National Action,* U.S. Government Printing Office, Washington, D.C.

TABLE 2.3

Marital Stability by Age Moved North, Age, Sex and Education (% of First Marriages Ending in Separation or Divorce)[a]

Sex, education, age	Age moved to the North		
	Born in North	Before age 10	After age 10
MALES			
High school diploma or more			
Age 30–45	33 (171)	22 (50)	26 (163)
Age 21–29	26 (99)	14 (43)	9 (95)
Without high school diploma			
Age 30–45	44 (110)	41 (69)	35 (485)
Age 21–29	20 (50)	21 (29)	9 (100)
FEMALES			
High school diploma or more			
Age 30–45	37 (167)	48 (85)	40 (290)
Age 21–29	28 (100)	22 (50)	19 (211)
Without high school diploma			
Age 30–45	56 (147)	43 (69)	44 (531)
Age 21–29	41 (124)	38 (80)	26 (112)

[a]Numbers in parentheses are figures on which percentages were based.

are older than their wives, and thus many divorced women in their twenties have ex-husbands who are in their thirties. But for men and women over 30, this age difference becomes less important, and it looks like a reporting bias; either never-married mothers are reporting a previous marriage, or divorced men are saying they have never married.

Table 2.3 also shows that among northern-born respondents, older women have only moderately higher rates of marital breakups than young women, despite the fact that they have an additional decade in which their marriages could dissolve. This means that the group of northern-born women now in their twenties will most certainly have a higher rate of marital breakup than the older women. Once again being born in the North increases the likelihood of a problem, in this case, marital instability, and the problem is growing worse. This is in contrast to the pattern for whites, for the divorce rate of whites remained constant from the end of the World War II to the early 1960's.[10]

The general attitude of our respondents toward marriage was cynicism. Twenty-eight percent of men and women who had been married said "no" when asked, "If you had your life to live over again would you marry or not?" Sixty-

[10] James L. Price (1969), *Social Facts* p. 173, Macmillan, New York. The divorce rate for whites has been increasing sharply since 1963. (See the Statistical Abstract for more recent figures than those given in Price.)

eight percent disagreed with the statement, "Most men make good husbands."
Fifty-nine percent said the same about women; the comparable numbers of
whites are 41% and 37%.

Black children are much more often involved in divorce than are whites. Our
respondents are also more likely to be orphaned than white children, since the
death rate for blacks in their thirties and forties when our respondents were
growing up was much higher than for whites. Forty-three percent of the respondents did not live with both their parents, and separation and divorce are the
primary causes. In contrast, only 11% of whites come from broken homes, and
most of these are broken by death, as only 2% of the whites had divorced or
separated parents.

This, then, is the pathology of the northern ghetto—poor jobs, low income,
high arrest rates, and interpersonal tension which produces fist fights and divorce.
In some ways things are getting better: nonwhite levels of educational achievement are increasing dramatically, and the black-white income gap is not as great
as it was. But in some ways, things are getting worse: marital instability is increasing and the reports of fighting and being arrested are discouraging.

Black Views of Their Situation

Blacks generally feel they have been victims of white racism in some form.
Fifty-four percent of the men in our sample can recall a childhood incident in
which a white adult subjected them to ridicule, ordered them to the back of a
bus, or did worse. Thirty percent of our respondents claim to have been denied
a job because of their race, and 23% can name an employer who they think does
not hire blacks. These numbers cannot be taken at face value since it seems
evident that a large number of blacks exaggerate in blaming whites for their
troubles, and many others are ostrich-like in their refusal to admit that they have
been victimized. So the real extent of mistreatment could be considerably lower
or higher than these percentages indicate.

It has often been suggested that much of the existing segregation in housing
is the result of self-segregation and the unwillingness of blacks to "pioneer" into
a white neighborhood. Our respondents, however, do not support this theory.
One in eight (13%) stated that he had been refused housing because he was black.
This figure seems valid, for in each case the respondent was asked about the incident; usually he was able to describe being told that an apartment was rented
when in fact the newspaper ad ran for days afterward, or could recall being told
quite bluntly that blacks were not permitted to live there. Thus it is not surprising that 70% of our sample describe their neighborhoods as "mostly black"
or "all black."

Racism, violence, and marital problems are all acts of aggression—the inflict-
ing of pain on others. We have measured three of the ways in which people hurt

each other and there are many others. Perhaps the most sobering statistic in the survey captures the way in which interpersonal relations in the ghetto are laced with the infliction of pain. Seventy percent of urban northern whites say "yes" to the statement, "Most people can be trusted." Only 32% of blacks agree, less than half as many.[11]

This underlying distrust of blacks and whites alike is evident in the questionnaires. A third (32%) of the respondents agree with the statement, "Sometimes I would like to get even with white people for all they have done to the Negro." Forty-two percent say that when they are around whites they are sometimes "afraid I might lose my temper at something he says." But at the same time, 46% say that sometimes "I am very careful not to make a bad impression." Many of our respondents seem to be saying that they are caught in a bind, trying simultaneously to make a "good impression" and to control their hatred of whites. Blacks should be uncomfortable, given the small number who actually have close white friends. Only 34% have no white friends, but more important, only 16% see their white friends "frequently." Even more disturbing are the feelings our respondents express about other blacks; 41% agree that "Generally speaking, a lot of Negroes are lazy." At the same time, 77% agree that too many blacks who have college degrees don't want to have anything to do with blacks who are not as fortunate as they are.

The respondent seems to see himself as an average person, suffering at the hands of the poor and lazy blacks who make a bad impression for the race, while the well-off ignore their plight and are prejudiced against the common man. (The way the items are worded may have unintentionally exaggerated the number of antiblack responses. For example, some respondents may feel that if there is one college-educated black who is not helping his brothers, that's too many. But even considering this, 77% seems large.)

What does all this add up to? If we had to choose one statistic, perhaps the most valid simple measurement of "success in living" is simply whether a person is happy or not. When whites are asked, "taking all things together, how would you say things are these days? Would you say you are very happy, pretty happy, or not too happy?" approximately 6% of all whites answer "Not too happy," a gentle euphemism for unhappiness. In contrast, 22% of blacks are "Not too

---

[11] One may raise the question of whether blacks are referring here to distrust for other blacks or whether they simply mean that most people are whites, and whites can't be trusted. In general it seems when race is not specifically mentioned most respondents assume other people to mean other blacks. In this case, there is a relatively low association ($Q = .09$) between response to this question and the response to, "If a black is wise he will think twice before he trusts a white man as far as he would another black." This low association indicates that the two items are *not* measuring the same thing, and therefore this item does not refer to whites.

happy," nearly four times as many. This is the largest white-black difference in the study, and we will try to explain it in Chapter 7.

Thus the portrait of the shattered dream concludes. There does indeed seem to be a seamless web in which poverty, discrimination, aggression, broken homes, distrust, and unhappiness are fused into a single unit. But we will not simply throw up our hands at "seamless webs" or "vicious cycles." Rather, we will try to dissect some of the strands in the web, and show how they are, in fact, causes and effects of each other.

Summary

Compared to the South, the North is a land of opportunity. Educational opportunities are far superior, and the industrialized North has less discrimination in employment. But the North has not been the land of opportunity worthy of the American dream of assimilation and equality. Life is getting better in the North, but slowly.

Poverty persists despite claims that each new generation of blacks is better off than the last and that blacks will soon be assimilated into American society much as other immigrant groups have been. The truth is that second-generation northern blacks are no better off, relative to whites, than are the migrants from the South themselves. Urbanization has meant only the opportunity to live in huge ghettos of poverty, violence, marital instability, and fear, without escaping discrimination and unhappiness.

3

A Theory of Black Poverty

Whites are the power structure of this country so they are definitely responsible for all Negroes' problems.[1]

In Chapter 2 we tested and rejected one theory of the failure of northern ghettos; the "urban immigrant" theory. When we look for other theoretical approaches to black poverty, we find little which can be considered a genuine theoretical approach. The problem is so complex that few writers have dared to handle the whole topic with a single theoretical argument.

But of course theories of black poverty do exist. Each slogan implies a theory; every program to alleviate poverty has a set of assumptions on which it is based. Let us briefly look at two programs designed to attack black poverty and ask what theory underlies each. First, we will examine the "war on poverty." Then we will look at the strategies of the civil rights movement. Last, we will present the social-psychological theory which we plan to test in this study.

The Poverty Program: Education, Social Work, Money

Of course any policy is created within a political reality, and often the politics are more important than anything else. This is certainly the case with the poverty program; by redefining the issue of racial inequality as an issue of poverty, the Johnson administration succeeded in making a politically unpalatable issue into one which was acceptable to all. Thus there are good political reasons why race

[1] From the questionnaire of a 22-year-old man, native of Minneapolis.

does not enter explicitly into the design of the poverty program. But what are the implications of such a policy? With what theory of race is it consistent?

In a nutshell, the poverty program committed itself to providing education, social services, and indirectly, money to the poor. A major emphasis of the program was on education: Head Start, job retraining, and the Job Corps. All three programs were built around the idea that a lack of either ability to learn in elementary school, or specific skills necessary for employment, or a high school diploma were the critical factors. Preschoolers were given the chance to get a "head start;" unemployed workers were given new job skills; and high school dropouts were taken off the street and encouraged to go back to school.

A second aim was to improve social services in the ghetto through the coordination of existing social work programs, development of new programs, or the reemphasis of existing programs away from middle-class clientele toward the poor. And finally, at the same time that the poverty program was being developed, other government officials were concerned with increasing welfare payments and forcing the economy toward full employment.

There is an implicit theory of black poverty underlying this set of priorities. It is that blacks differ from whites in that they are poorer, less skilled, and lack adequate educational opportunities. If blacks could be somehow pulled out of poverty and given enough education, they would manage to make the rest of the jump into the economy on their own. The poverty program can be viewed as a "conservative" solution to poverty because it stresses change in the individual rather than the system. But the theoretical argument underlying the program seems unsatisfactory. First, the emphasis on education seems misplaced. Why, for example, should blacks need vocational education, when most whites have had nothing more than on-the-job training?

The assumption is that blacks have poor education because they lack good schools and because they come from poor homes where they are poorly prepared for school. Granted, educational opportunities in northern ghettos are not equal to those in good white schools, but on the other hand, they are far superior to those in southern black schools. Alumni of southern black schools who come North score below native Northerners on intelligence tests, but we have already seen that they have higher incomes.

In addition, the Coleman report[2] was unable to show that differences in educational opportunities *caused* the low achievement of blacks; the best that could be said is that ghetto schools did nothing to help blacks catch up to whites.

The emphasis on inadequate home preparation for school, and the emphasis on social services, suggest a "culture of poverty" theory; that poor people, re-

[2] James S. Coleman, Ernest Q. Campbell, Carol J. Hobson, James McPartland, Alexander M. Mood, Frederic D. Weinfeld, and Robert L. York (1966), *Equality of Educational Opportunity,* U.S. Office of Education, Washington, D.C.

gardless of race, have a special culture—a set of rules for living, customs, and values—which prevent achievement. However, one review of the literature[3] concludes that the poor do not have a separate culture; what the poor want from life, what they believe they should do, and what they believe to be right and wrong do not differ greatly from what the middle class believes. (Witness the commitment of the poor to education.)

The poverty program has recognized (we think correctly) that poor blacks have many difficulties in managing their personal lives and their relationships with other people. But the poverty program lacks a defensible theory of why these problems exist. For example, the poverty program emphasizes education, even though the steady increase in educational opportunities since the end of World War II seems to have had little effect on the black-white income gap. And there is little reason to believe that blacks need to be acculturated, i.e., taught to change their values. The improvement of social services, like improved education, seems a worthwhile goal, but it is hard to see how this can be more than a "band-aid" approach to the problem. Similarly, job training programs may be useful, but it is hard to believe that the man with a criminal record has that record primarily because he lacked a job skill, or that providing him with training will solve his problems. In short, the war on poverty might work, if blacks were indeed a new immigrant group with low literacy and no skills. But they are not; and this strongly suggests that something in the nature of the society has prevented blacks from taking advantage of the education and skills they do have.

Any theory of black poverty must recognize that the black and white poor are different in their adaptations to the different societal constraints placed upon them. One reason why the poverty program lacks such a theory is that political considerations have prevented the program from citing racial explanations (e.g., the racism of specific institutions) or explanations couched in terms of the failure of the social structure. The argument for "community control" as adopted by the poverty program[4] does recognize that the poor do not control their environment, and in this sense the program can be said to recognize inherent problems of the social structure; but, in fact, the poverty program is not committed to any change in the structure of society beyond increasing the participation of the poor in existing institutions.

We do not mean to imply, however, that discrimination or institutional racism account completely for the amount and persistence of black poverty. We must

---

[3]Zahava D. Blum and Peter H. Rossi (March, 1968), "Social Class Research and Images of the Poor: A Bibliographic Review," The Johns Hopkins University, Center for the Study of Social Organization of Schools, Baltimore.

[4]Peter Marris and Martin Rein (1967), *Dilemmas of Social Reform: Poverty and Community Action in the United States*, Atherton Press, New York.

also recognize that blacks and whites are subject to different social expectations and employ different adaptations to their environments, some of which may be dysfunctional in terms of reconciling the races. Undoubtedly there is a great deal of racism serving to "keep the black man down," but there is also a good deal of personal disorganization that is unaffected by superficial training programs and prevents blacks from achieving. Problems of alienation, lack of assertiveness or middle-class ambition, alcoholism, or low I. Q. are not racist myths that disappear on command. These problems may have had their ultimate cause in the systematic exclusion of blacks from full citizenship over the past two centuries, but this does not mean that removing the discriminatory barriers in employment will automatically bring blacks out of poverty.

The personality traits and styles of behavior which are statistically more common among blacks than among whites and which serve to prevent assimilation or economic progress include: (1) high rates of violence and crime; (2) high rate of escapism, including use of drugs and alcohol; (3) excessive apathy and timidity; this appears in the difficulty experienced in recruiting blacks for job openings, or for newly integrated housing, in registering black voters, etc. (This characteristic has also been called "distrust of the system"); (4) low learning ability, in both academic and job situations; this is usually demonstrated by pointing to the generally lower I. Q. scores of blacks; (5) high levels of interpersonal difficulty, especially in marriage; this is apparent in the higher proportion of broken marriages among blacks. The reader no doubt has an uneasy feeling about these points, but these are facts that must be dealt with in order to understand black poverty. These are not racist claims since we do not claim that the characteristics are necessarily genetically controlled or that they are unalterable even if social conditions are changed. It is more likely that these are culturally determined characteristics, adaptive responses to a given societal situation, and not tendencies inherent in the individual.

What we are arguing here is not very complicated, but it is difficult to understand (for the writers as well as the reader) because of the intense ideological conflict surrounding the issue of black poverty. At the one extreme, conservatives blindly deny that discrimination exists, or that it is of any importance. Here is a job opening, and no qualified black has come forward to fill it. Therefore "it's their own fault." At the other extreme, the white or black radical denies that blacks have any personal disabilities. It is all the fault of society; present-day white racism—meaning overt discrimination—is the only villain. Both positions are misguided, of course, and neither position leads to a reasonable solution. The conservative program, that blacks should change themselves, become more virtuous, and then they will receive their just desserts, is wrongheaded for two reasons. First, there really is economic discrimination in employment, in the cost of consumer goods, and the cost of housing, which no amount of personal virtue can overcome. Second, how do we help people to change, to become more virtuous, when we have no notion of why they are the way they are? This line of reasoning leads ultimately to the notion that the incompetence of "the poor" (meaning blacks) is immutable.[5]

Unfortunately, the radical argument on the other side is not much more helpful. It denies that blacks have faults. In its extreme form, one may hear it argued that black crime is reasonable behavior on the part of the criminal, and can be justified as a useful political response to white racism. But in fact, if all discrimination in employment were eliminated immediately, personal disorganization would still prevent many blacks from making a go of it. The question of why black personal disorganization exists is not answered by conservatives and not asked by radicals.

The conventional wisdom of both the conservatives and the radicals agrees with the thesis of the poverty program: blacks are not in any important way different from whites. The conservatives are partly trying to conceal their racial prejudice by claiming to speak about all the poor, both white and black. In part, they are denying that blacks are different from whites in order to deny that discrimination exists. The radicals, on the other hand, deny that blacks are different from whites because of the fear that these differences may be genetically controlled, or may in some way mean that blacks as a group are morally inferior to whites. Blacks are oppressed, but only in highly tangible ways such as employment discrimination. In this ideological climate, everyone benefits by redefining the issue as a concern with "the poor." Low socioeconomic status becomes the explanation for everything, and sometimes, in a completely circular fashion, for poverty itself.

The Civil Rights Movement: Nonviolent Integration and Brotherhood

In many ways, the assumptions underlying the civil rights movement seem almost opposite to those of the "war on poverty." We can see this in the writings and work of Martin Luther King, Jr., which show, first, less of a concern with real issues, such as jobs or welfare payments, than with the symbols of racial inequality. He seemed to believe that the symbols must be removed before the real issues could be addressed. Second, there is a consistent focus upon nonviolence through the use of law and passive civil disobedience. Third, there is a shutting off of aggression toward whites and a stress on solidarity with whites who are sympathetic to the cause. Fourth, stress is placed on religious values and the virtue of hard work, and fifth, there is a belief that the ultimate goal of the movement is brotherhood between individual whites and blacks.

[5] The best known source of this argument is Edward C. Banfield (1968), *The Unheavenly City*, Little, Brown, Boston.

Like the poverty program, this is a political strategy as much as it is a technique for racial advancement. As a strategy for achievement it seems beset by contradiction. But King's own writings point to his recognition of these apparent dilemmas. On the reconciliation of brotherhood with aggressive demonstrations aimed at integration King said,

> ... I do not think that violence and hatred can solve this problem. They would end up creating many more social problems than they solve. I'm thinking of a very strong love, of love *and* action, not something where you say, "Love your enemies," and leave it at that. But you must love your enemies to the point that you're willing to sit at a lunch counter in order to help them find themselves.[6]

When the Supreme Court ruled that segregation in transportation was in violation of the Constitution, thus ending the Montgomery bus boycott, King issued the following statement:

> Our experience and growth during this past year of united non-violent protest has been such that we cannot be satisfied with a court "victory" over our white brothers. We must respond to the decision with an understanding of those who have oppressed us and with an appreciation of the new adjustments that the court order poses for them. We must be able to face up honestly to our own shortcomings. We must act in such a way as to make possible a coming together of white people and colored people on the basis of a real harmony of interests and understanding. We seek an integration based on mutual respect.[7]

Why is it necessary to love whites at the same time that you exert the greatest possible pressure on them? Perhaps because King sees nothing less than the love of blacks and whites for each other as the true goal. Herbert Storing[8] builds a conservative critique of King around a critical sentence from *Stride Toward Freedom*.[9] Storing first quotes Justice Brown in the famous Supreme Court case of *Plessy vs. Ferguson* in which the court found segregation in railroad transportation legal: "If enforced separation is regarded as stamping the colored race with a badge of inferiority, it is not by reason of anything found in the act, but solely because the colored race chooses to put that construction upon it . . ." Storing then adds, "The higher truth of this statement may be seen by comparing this to the reaction to segregated travel of Martin Luther King. . ." and quotes King's feelings the first time he was seated behind the curtain in a dining car:

[6] Robert Penn Warren (1965), *Who Speaks for the Negro?* p. 211, Random House, New York.

[7] *Ibid.*, p. 205.

[8] Herbert J. Storing, "The School of Slavery," in *100 Years of Emancipation* (1963), p. 68, (Robert A. Goldwin, ed.) Rand-McNally, Chicago.

[9] Martin Luther King, Jr. (1958), *Stride Toward Freedom,* Harper & Row, New York.

"I felt as if the curtain had been dropped on my selfhood." Storing writes "He could never 'adjust' to separate accommodations not only because the separation was always unequal, but 'because the very idea of separation did something to my sense of dignity and self-respect.'" Storing then compares this to Frederick Douglass commenting on being forced to ride in a baggage car in Pennsylvania: "Mr. Douglass straightened himself up on the box which he was sitting on, and replied: 'They cannot degrade Frederick Douglass. The soul that is within me no man can degrade. . .'"[10] Thus Storing correctly focuses upon a critical point in this disagreement between King and Douglass. King is psychologically wounded by segregation; Douglass pays no attention. The question is, who is the wiser psychologist, Douglass or King? Is King's reaction pathological or is Douglass'? For that matter, which court decision displayed the greater understanding of the psychology of race, *Plessy vs. Ferguson* or *Brown vs. Board of Education,* which found school segregation to be psychologically harmful? In short, how should a normal man react to discrimination?

In most of King's writings there is a tenor of the restraint of the Christian rhetoric, but when King talks about the psychological consequences of segregation his writing seems to become more strident. Consider the following two quotations:

> The absence of freedom imposes restraint on my deliberations as to what I shall do, where I shall live or the kind of task I shall pursue. I am robbed of the basic quality of manness. When I cannot choose what I shall do or where I shall live, it means in fact that someone or some system has already made these decisions for me, and I am reduced to an animal. Then the only resemblance I have to a man is in my motor responses and functions. I cannot adequately assume responsibility as a person because I have been made the victim of a decision in which I played no part.
>
> This is why segregation has wreaked havoc with the Negro. It is sometimes difficult to determine which are the deepest wounds, the physical or the psychological. Only a Negro understands the social leprosy that segregation inflicts upon him. Like a nagging ailment, it follows his every activity, leaving him tormented by day and haunted by night. The suppressed fears and resentments and the expressed anxieties and sensitivities make each day a life of turmoil. Every confrontation with the restrictions against him is another emotional battle in a never-ending war. Nothing can be more diabolical than a deliberate attempt to destroy in any man his will to be a man and to withhold from him that something which constitutes his true essence.[11]

> Being a Negro in America means not only living with the consequences of a past of slavery and family disorganization, but facing this very day the pangs of "color shock." Because the society, with unmitigated cruelty, has made the Negro's color

[10]Storing, *Ibid.*, p. 100.
[11]Martin Luther King, Jr. (1967), *Where Do We Go From Here, Chaos or Community*? pp. 99, 109-110, Harper & Row, New York.

anathema, every Negro child suffers a traumatic emotional burden when he encounters the reality of his black skin. In Dr. Kenneth Clark's classic study, *Prejudice and Your Child*, there is a simple test for color sensitivity in young children. This involves having a child draw a tree, an apple, and a child, and then proceed to color each one at will. One of my colleagues reports conducting this test with his three-year-old daughter. The child, holding the crayon deftly and delicately, colored the tree green and then in the same way shaded the apple red. But when it came to coloring the child, she gripped the crayon with her fist and in a violent pattern of chaotic motions made purple slashes across the figure.

. . . Every Negro comes face to face with his color shock, and it constitutes a major emotional crisis. It is accompanied by a sort of fatiguing, wearisome hopelessness. If one is rejected because he is uneducated, he can at least be consoled by the fact that it may be possible for him to get an education. If one is rejected because he is low on the economic ladder, he can at least dream of the day that he will rise from his dungeon of economic deprivation. If one is rejected because he speaks with an accent, he can at least, if he desires, work to bring his speech in line with the dominant group. If, however, one is rejected because of his color, he must face the anguishing fact that he is being rejected because of something in himself that cannot be changed. All prejudice is evil, but the prejudice that rejects a man because of the color of his skin is the most despicable expression of man's inhumanity to man.[12]

The assumption then is that dignity is something which blacks cannot attain for themselves in a segregated society except perhaps through their commitment to a future in which brotherhood would exist. Dignity is something that whites have taken away from blacks and only whites can give back to them.

Here we see the beginning of a theory of black poverty which argues that in some way segregation robs the black of some vital aspect of his personality, which in turn prevents him from realizing his ambitions. We have argued that blacks have more personal problems than whites. If King is right, and we can show that segregation is the primary cause of this potpourri of difficulties that blacks have, then we will have succeeded in producing a theory which does synthesize the fact of discrimination with the fact of black disability.

A Social Psychological Theory

The idea that discrimination causes black personality problems is hardly new. Indeed, much of what we have said was written by Gunnar Myrdal in the 1940's.[13] The most comprehensive single statement of the theory, written in 1951, is *The Mark of Oppression,* by Abram Kardiner and Lionel Ovesey.[14] Kardiner was a member of the "culture and personality" school, which sought to unite anthro-

---

[12]*Ibid.*, p. 100.
[13]Gunnar Myrdal (1944), *An American Dilemma,* Harper & Row, New York.
[14]Abram Kardiner and Lionel Ovesey (1951), *The Mark of Oppression,* Norton, New York.

pology and Freudian psychology, tracing out linkages between characteristics of a particular culture, such as toilet-training customs, and adult behavior in that culture. Much of this writing was speculative, frequently with no real data on the personality traits which supposedly derive from the culture. In an effort to provide better personality data, Kardiner enlisted Ovesey to carry out psychiatric diagnoses of 25 black subjects, including both psychiatric patients and paid subjects.

Scattered regularly throughout most of the cases are vivid black and white images—dreams of being chased by men in white clothes or fantastic self-images of black rats. But no simple racial theme appears in the interpretation of the dreams and free associations of the subjects. Instead, the traditional psychodynamic analysis of parent-child relationships are used to explain the behavior of the subjects. Of course, analysis of the family is the only theoretical approach psychiatry has developed so far. We have developed the techniques to observe the familial roots of psychopathology, but lack the tools to study the impact of the larger society on the personality. (We have, for example, no theory with which to interpret the racial references in dreams.) Even if the data were clearer, however, a sociologist would be reluctant to take seriously a sample of 25 self-selected or neurotic people as in any way representative of blacks in general.

Despite these problems, Kardiner and Ovesey do develop the beginnings of a theory of "the mark of oppression." The first central relationship is that discrimination depresses black self-esteem. The individual cannot avoid internalizing the low opinion whites hold of him and he feels worthless, disliked, and inadequate. He must then go about finding defenses to protect himself from the surfacing of this unconscious sense of badness. But the defenses available to him are influenced by a second key factor. Being a victim of poverty, emotional deprivation, and discrimination has left the subject frustrated and angry. Following Dollard,[15] Kardiner and Ovesey argue that blacks have learned to defend against their aggressive needs by sharply internalizing them. They do not express anger at others at all, but they do become angry at themselves, either consciously or unconsciously. Thus their already low self-esteem is further depressed by high levels of aggression directed at themselves. The black who, at the unconscious level, sees himself as black and worthless, compounds this with great quantities of recrimination and self-blame.

According to psychodynamic theory, aggression is inhibited in part because it is frightening: the subject is fearful of what he might do. One patient, being treated for sexual impotency said, "Very seldom do I get angry. It takes two to make an argument, and I don't argue. I just don't talk anymore. I just lay down and read a book. I don't do anything. I guess I never been angry more than two or three times in my whole life that I can remember. Lots of times I think if I get

---

[15]John Dollard (1957), *Caste and Class in a Southern Town,* Doubleday, Garden City.

pushed to a certain point, I'll get mad, but it never gets to that point. I often
wonder if I get angry, what would I do."[16] Actually the subject, it turns out,
knows something about "what he would do." He once lost his temper, grabbed
a pistol, and fired it at a fellow soldier. As it happened, the pistol was unloaded.
In his case, his sexual difficulty is linked to his subconscious fear of attacking a
woman—hurting her through intercourse—and being hurt by her in retaliation.
Being unable to control his anger when he does express it, he suppresses his
anger, letting it appear in such indirect ways as impotency.

There is no explicit reference to race in the analysis of this subject; a white
person with strong needs to express aggression could react in the same way. But
Kardiner argues that for blacks inhibition of aggression is common because
blacks must passively play the role of inferiors to whites, who have enormous
power over them.[17] Similarly, Myrdal notes that whites are very conscious of
black aggressiveness, but, "In view of the fact that they are so frequently discriminated against and insulted, Negroes are remarkably passive and polite
towards whites."[18]

Rorschach tests were administered to 24 of the Kardiner-Ovesey subjects by
William Goldfarb, who interpreted them as showing high levels of rigidly controlled aggression. In particular, 21 of the 24 tests offer at least one image of
human physical mutilation. This may be simply another expression of aggression,
but Kardiner notes the similarity of this result to the symptoms of war psychosis
and argues that it reflects a lack of control over the subject's world. It seems
plausible that some of these subjects are identifying themselves with the mu-
tilated image. Kardiner notes that,

> The techniques for disposing of rage are varied. The simplest disposition is to sup-
> press it and replace it with another emotional attitude—submission or compliance.
> The greater the rage, the more abject the submission. Thus, scraping and bowing,
> compliance and ingratiation may actually be indicators of suppressed rage and sus-
> tained hatred. Rage can be kept under control by replaced with an attenuated
> but sustained feeling—resentment. It may be kept under control, but ineffectively,
> and show itself in irritability. It may be kept under sustained control for long
> periods, and then become explosive.[19]

If these were the only manifestations of suppressed rage, the analysis would be
straightforward. Unfortunately, as Kardiner goes on to argue, unexpressed aggression can take a number of other forms:

---

[16]Kardiner and Ovesey, *op. cit.*, pp. 94-95.
[17]*Ibid.*, pp. 303-304.
[18]Myrdal, *op. cit.*, p. 958.
[19]Kardiner and Ovesey, *op. cit.*, pp. 304-305.

> Rage may show itself in subtle forms of ingratiation for purposes of exploitation. It may finally be denied altogether (by an automatic process) and replaced by an entirely different kind of expression, like laughter, gaiety, or flippancy.[20]

The most striking example (a paid subject, not a patient) was a porter whose livelihood depended on manipulating whites for tips. Here he talks about the white drunks he dealt with as a washroom attendant:

> That's what they always do—start picking on a colored guy. That's the first thing these hill-billies do when they get drunk. Well, I tries to take them back to the wash basin and get them to clean up. I tells them, "Come on back here and freshen up." Then I tries to get them to take a towel. They says, "How much is it worth?" Well, it's worth five cents, but I smiles at them and says, "Whatever you feels it's worth, boss." I shame him out. It's better that way. You understand? (You put yourself in the lower position.) Yes, I do that. (How does this make you feel?) Well, you just say words ain't going to hurt you, especially when you know you get paid for it. I had a Captain came in one Christmas time. He gave me five dollars to shine his shoes five times. He was drunk and had his bottle with him. He said, "I'm going to give you a dollar every time you shine my shoes and I'm going to call you old nigger." And he did. (Here the subject laughs. He goes on:) It hurt me, sure, but it was the only way to get his money and get rid of this guy. So let him talk.[21]

Unfortunately, he was unable to modify the "boy" behavior when he was out of the washroom. Later he attempted to borrow $100 from the psychiatrist; when he was refused, he dreamt that Ovesey only paid him 10¢ per hour for coming to the sessions but he was very pleased to receive the dime. Ovesey probed hard at the meaning of the dream, but the subject refused to say that he was angry at the refusal of the loan:

> (I think this dream means you were mad because I wouldn't give you the $100 and you're covering up by laughing.) Roars with laughter. "Mad!" Naw, I ain't mad" (You're saying in this dream, "Dr. Ovesey is a cheap bastard," but you're covering up with laughter.) Laughs almost hysterically, slapping his thighs with his hands. "No, it's like this: $100 is a lot of money. You don't just walk up to a man and ask him for $100. No, I wasn't mad. If I was that mad, it would have come out and I would tell you." (You did tell me. It came out in the dream.) Laughs uproariously. "That ain't the way I feel. I don't see it in my mind." Laughs and laughs and laughs. (You don't like to tell me.) "You know if it was there, Doc, I would tell you. I couldn't walk around with it on my chest. It would have come out."[22]

---

[20]*Ibid.*, pp. 304-305.
[21]*Ibid.*, p. 103.
[22]*Ibid.*, p. 106.

All of these devices for deflecting aggression, Kardiner argues, mean that there are fewer realistic ways to secure self-esteem. Rather, the subject sets unrealistically high aspirations, does not follow through, becomes anxious, and ultimately develops a passive demeanor and a general inhibition of affect.

Aggression is then one of the two central components of Kardiner's theory. Blacks, because they are discriminated against, are angry; because they must be subservient, they deny or deflect this aggression. The other central concept is self-esteem. Discrimination lowers self-esteem, and since the subject cannot aggressively battle whites to regain that self-esteem, he must look to other sources to bolster his damaged sense of worth. Together these factors produce: (1) depression and anxiety from internalized aggression; (2) violent outbursts when (and if) the aggression is released; (3) compensatory efforts to recapture self-esteem through a highly stylized manner of dress and behavior; (4) passivity, goallessness, and "laziness." and (5) inordinate demands for gratification placed on friends and family, leading to interpersonal tension.

Kardiner's argument successfully ties the passivity of black spiritual music to the violence of Chicago's 63rd Street, the laughing manner of the southern slave to the sullen depression of the northern militant.

This view of black personality is apparently a highly controversial one. It seems implicit in the work of Harold Finestone[23] and William McCord et al.[24] On the other hand, the best recent reader on race[25] makes no reference to this approach at all. To some social scientists, it is taboo to speak of black incapacities, even if one carefully softens the accusation by blaming these incapacities on white racism. Lee Rainwater[26] has focused directly on the equation we are concerned with: that white racism causes black incapacity which causes poverty. However, he concludes that the remedy lies in increasing black income by increases in either earned income or welfare—this would probably not be King's recommendation.

If King and Kardiner are right, the remedy for black poverty is to eliminate not merely the real socioeconomic status difference between blacks and whites, but the symbolic status differences as well.

Plan of the Book

In order to test the King-Kardiner hypothesis, we will first attempt to show in Chapters 4-7 that blacks do have personal difficulties which hinder their achieve-

[23]Harold Finestone, (1957), "Cats, Kicks, and Color," Social Problems, 5, 3-13.
[24]William McCord, John Howard, Bernard Friedberg, and Edwin Harwood (1969), Life Styles In the Black Ghetto, Norton, New York.
[25]Norval D. Glenn and Charles M. Bonjean, (eds.) (1969), Blacks in the United States, Chandler, San Francisco.
[26]Lee Rainwater, (1970) Behind Ghetto Walls: Life In a Federal Slum, Aldine, Chicago.

ment. In Chapter 4 we will see that blacks do have the kind of problems with aggression that Kardiner describes. In Chapter 5 we will show that blacks are more fatalistic than whites; they do not have as strong a sense of being able to control their lives. In Chapter 6 we will look at black self-esteem, although here our data do not agree with Kardiner. Chapter 7 ponders the fact that blacks are more likely than whites to say they are unhappy. Chapter 8 is a summary of the preceding four chapters and shows that these personality traits do seem to work to prevent achievement.

In Chapters 9 and 10 we will attempt to explain some of those personality difficulties, first by examining the impact of the unstable family, then the impact of racial discrimination. In Chapter 11 we will take a more careful look at one type of racial inequality: racial segregation in schools. Chapter 12 concludes with a discussion of the implication of these data for social policy.

4

Aggression

My grandmother told me to stay out of the neighborhood. A neighbor's children, a white neighbor, came down to my yard and told me to stick my hand through a picket fence and they would give me some money. I put my hand through and the two white brats cut my hand with glass. My hands were very badly cut. (He showed interviewer deep scars on his right hand.) My grandmother told us to just stay out of their company or she would tear my behind up. She told me just to ignore white people.[1]

This quote is from a respondent for whom racism has a vivid meaning. One can easily imagine him bearing an enormous grudge against a world where white children attack him with impunity and his grandmother seems to join in the assault rather than protect him. But this is an extreme case. Are the lives of most people touched directly by racism, or is racism merely an intangible set of rules which need no enforcement and carry no more sense of cruelty than, say, the Blue laws which prevent many of us from Sunday shopping? This is an important question, for the argument that blacks are saddled with unmanageable levels of internal aggression hinges on how vivid and painful racial discrimination is.

It is interesting to speculate about the reasons for social scientists' preoccupation with aggression among blacks. Even before urban crime statistics revealed disproportionate rates for the black ghetto or the riots of the sixties had occurred, the study of black aggression had made some of its classic statements about the various ways in which blacks channel the aggressions assumed to be aroused by

[1] Questionnaire 1180, a 26-year-old man who grew up in New Orleans.

interracial tension.[2] Probably some of this interest was due to the novelty of
employing psychoanalytic hypotheses about the functions of "defense mechanisms" in anthropological fieldwork as well as to the white investigators' conscious or unconscious acceptance of prevailing black stereotypes. And possibly
out of guilt, the assumption is usually made that there is a higher "level" of
aggression in the black community due to the oppression of racism and that
blacks therefore are prone to express more aggression in socially dysfunctional
ways. For whatever reasons they were made, many of the assumptions about the
origin and handling of aggression among blacks have yet to be demonstrated in
a representative sample of the adult black population.[3] Even the high correlation
between incidence of crime and the blackness of a population says nothing about
the generation of aggressive impulses by racism or about the relationship of
modes of aggression-handling to racial variables. Furthermore, the psychological
mechanisms whereby aggressive impulses are expressed, deflected, or suppressed
must be common to all races, yet no studies have attempted to examine these
mechanisms for comparable samples of blacks and whites. This chapter will
document the effects of experiences of racism and discrimination on expressions
of aggression among our respondents and the effects of various types of aggression-handling on respondents' behavior and racial attitudes. Later chapters will
develop the black-white comparison further.

From our questionnaires it does appear that most blacks feel they have been
personally touched by discrimination. Forty-five percent of blacks recall childhood mistreatment by white adults. Some are more severe, but most less serious,
than the incident described above.[4]

Add to this the 13% of our respondents who say that they personally have
been refused housing (and discrimination in housing is usually so blatant that we
have relatively little reason to believe they are exaggerating) and it is surprising
that there are so many instances of discrimination reported. One would assume
that it should be clear where blacks are or are not wanted, and it should be easy

[2] Hortense Powdermaker (1943), "The Channeling of Negro Aggression by the Cultural
Process," *Am. J. Soc.* **48,** 750-758.

[3] There have been several studies of children, few of adults, and those relating to adults
usually are psychoanalytic case studies of a small number of people. See, for example:
William H. Grier and Price M. Cobbs (1968), *Black Rage,* Bantam Books, New York; Herbert
Hendin (1969), *Black Suicide,* Basic Books, New York; and Kardiner and Ovesey (1951), *The
Mark of Oppression,* Norton, New York.

[4] Actually, the reported incident is unusual, since the attacking whites are children rather
than adults as called for by the question. Also, the questionnaire does not ask the respondent
to describe the incident but merely to say whether he can recall one or not. One reason why
the question was shortened was that after the pretest a charming and excellent interviewer
who is almost a stereotype of the "middle-class black," suggested it be dropped to save time,
"Otherwise we will just have to listen to all those stories."

to avoid applying for housing in the wrong neighborhood. In fact, however, this is not the case. Roughly a million blacks have personally met discrimination in housing. The large number of respondents who report experiencing discrimination in this one area suggests that a much larger number could report experiencing prejudice if we broadened the question to deal with everyday activities such as shopping, riding in taxis, dealing with public officials, etc.

Add to this the presence over the generations of a most ominous weapon—the lynching. During the lifetime of a 65-year-old man, there have been 1872 lynchings in the U.S.[5] This may seem a small number, but how many people have been indirectly touched by it? For each dead man there are at least scores, perhaps even hundreds of distant relatives who would hear of his death. Presumably neighbors would have watched from their houses as the night riders came, and many others in that area would know the victim by sight or by name. It is easy to imagine that these lynchings may have touched a million men, women, and children.

Blacks also experience frustration as an indirect result of discrimination: if unemployment causes a father to desert his family, the child may grow up angry at his father. Since blacks experience more hardships, they will be angrier than whites.

Finally, race provides a convenient scapegoat for blacks, just as it does for some whites. The black who cannot get a job is free to blame his failings on discrimination, and no one will dispute his opinion.

This leads us to our first point: between one's own experiences (and imagination) and those of friends and relatives, racism has had either a direct effect, an indirect effect, or an imagined effect, on the lives of most blacks now living in the North. This leads to our second assumption: at least to a small extent, every black adult carries with him the inexpressable need to retaliate. And, finally, to our third assumption: one of the major differences—perhaps *the* major difference —between blacks and whites is that blacks are forced to suppress a much larger amount of hostility. This is the hypothesis advanced by Kardiner and Ovesey, and also by Allison Davis, John Dollard, William Grier, and Brice Cobbs.[6]

Of course nearly every psychiatrist will argue that whites also must learn to control a great deal of aggression, and certainly a highly urban and industrial society provides little opportunity to "let off steam." Aggression is a problem for whites but we are arguing that it is a much more severe problem for blacks.

---

[5] Figures are compiled by the Tuskegee Institute and are widely reprinted. These are taken from Louis H. Masotti and Don R. Bowen, Eds. (1968), *Riots and Rebellion: Civil Violence in the Urban Community*, Sage Publications, Beverly Hills, California.

[6] John Dollard (1937), *Caste and Class in a Southern Town*, Yale Univ. Press, New Haven, Conneticut; William Grier and Brice Cobbs (1968), *Black Rage*, Basic Books, New York; and Allison Davis and John Dollard (1940), *Children of Bondage: The Personality Development of Negro Children in the Urban South*, American Council on Education, Washington, D.C.

We cannot make a precise prediction about the relationship of race to hand-ling of aggression. The work of Bertram Karon[7] strongly suggests that blacks do handle aggression differently from whites. He compares black and white re-sponses to a precoded variant of the TAT known as the Picture Apperception Test which had been administered to a national sample, and found that the pic-tures related to aggression consistently distinguished blacks from whites. But we cannot really predict the way in which blacks and whites will differ. Presumably the child who discovers that he is black, and therefore holds a low status, will be frustrated and angry. He can react by acting out his anger, displacing onto a safe victim (kicking his dog), or he can internalize it as Kardiner and Ovesey predict. We hypothesize that having learned one or more of these styles of aggression-handling, he will use it to handle nonracial frustrations as well. In adulthood, his behavior serves as a model for his children, and the pattern of aggression-handling becomes perpetuated.

We hypothesize that blacks will internalize aggression more than whites, but one could argue the opposite. It is true that blacks cannot act out their anger toward whites, but they can express anti-white feelings to other blacks, and thus they may be able to use "the man" as a scapegoat for their aggression. Such a person may find it easier to externalize aggression.

In our analysis of aggression, we will look at both types of aggression-hand-ling: first those that externalize, and then those who direct their anger inward.

Uncontrolled Aggression

It goes without saying that high rates of violence are a problem in the ghetto. More importantly, these rates are symptoms of widespread aggression which expresses itself in many different ways and hampers the black in many areas of his life. The same factors which lead a person to violence, for example, will also operate to lower his income. To make this point, let us first locate those respon-dents who have the most trouble controlling aggression. One group is those who answer "yes" to the question, "Have you ever been in a fight (not an argument, but a real fight) since you were an adult?" Twenty-eight percent of our respon-dents said yes, compared to 19% of whites. These respondents—the fighters—provide us with one group to examine. A related group are arrestees: 22% of the subjects say they have been arrested. This is much higher than the 9% rate for the sample of northern whites the same age.

We did not ask why they were arrested, but we feel safe in assuming that the typical arrestee has poor controls on his aggression, since arrestees are very likely to also say they have been in a fight. (We should note that an insignificant num-ber of these are arrests for civil rights activity. Twenty-two percent of the adult

---

[7]Bertram Karon (1958), *The Negro Personality,* Springer, New York.

black population is 3 million people, and the total number of civil rights arrests must be much smaller than this.)

The quality of these two questionnaire items is good in one sense, but poor in another. The items are good in that they are measures of actual behavior, rather than merely attitude questions. Presumably we can judge a person's personality better from what he does than from what he says. On the other hand, both items probably have high measurement error. We can expect large numbers of people to lie about being arrested; and people may differ in their definitions of "fight." Indeed the fact that persons who report being in a fight are also very likely to report being arrested increases our suspicion that there is a response bias. Some persons are much more willing to say "yes" to this type of question than others. Finally, we are really interested in the type of person who exhibits uncontrolled aggression, and not every such person has been arrested or in a fight, and conversely not every fighter or arrestee is aggression-prone. But let us use these two items, for lack of anything better, to identify that group of people whose controls on aggression are so low that in the face of a particularly frustrating situation, they come up swinging.

The Economic Consequences of Uncontrolled Aggression

The next question is, is the person who has weak controls on his aggression less likely to succeed? Obviously the person who has spent time in jail or been injured in a fight will suffer an economic loss as a result. But we are concerned with a subtler phenomenon: whether the person with weak controls on aggression has difficulties over and above the direct consequence of fighting or being arrested.

Only 38% of the arrestees are high school graduates compared to 56% of those who say they have never been arrested. This fact poses a problem we will encounter a number of times in the analysis: does being a high school graduate make one less prone to arrest, or is it weak control of aggression which prevented the respondent from finishing high school?

Our best guess is that both these statements are true. The poorly educated have neither the money to escape frustration nor the intellectual ability to rationalize it. On the other hand, the person with weak controls on his aggressive tendencies is likely to react to the frustrations at school by dropping out; his later arrest is caused not by his lack of education but by the personality trait that prevented him from completing school.[8]

---

[8] This view is supported by a Survey Research Center study which shows that "dropouts were above average in delinquency [before they quit school] . . . and there is no indication that this delinquency increased as a result of dropping out." See Jerald G. Bachman (March 1972)," Anti-dropout Campaigns and other Misanthropies" *Society,* 9 (No. 5), pp. 4, 6, and 60.

This general problem of which variable is cause and which is effect will make our analysis more complex. We wish to show that the person with weak controls on aggression is less successful economically. But of course poverty may "cause" the arrest, so simply showing that arrestees are poorer is unconvincing.

Table 4.1 shows that arrestees change jobs more often. What seems to happen is that the aggressive people in our sample run into trouble because they cannot tolerate the sort of frustrations most people must put up with in order to be reasonably successful. The fight- and arrest-prone respondents react to unpleasant, low-paying jobs where the payoffs are in such long-term factors as seniority, by simply quitting. The table is complicated because (1) we must look at men and women separately; (2) aggressive people have less education and consequently make less money (all poor people are more likely to change jobs, so we must control on income level); and (3) young people make less money and change jobs

TABLE 4.1

People Who Have Been Arrested are More Likely to Change Jobs (%)

| Number of jobs in last 5 years | Under age 29 | | | | Age 30–45 | | | |
| | Under $6000 | | Over $6000 | | Under $6000 | | Over $6000 | |
	Arrest	No arrest	Arrest	No arrest	Arrest	No arrest	Arrest	No arrest
Males								
1	6	33	29	27	40	50	76	69
2–4	54	59	53	68	44	45	23	30
5+	40	9	18	6	16	6	1	1
	100	99	100	101	100	101	100	100
N	(166)	(267)	(34)	(139)	(241)	(309)	(147)	(298)
	$\gamma = .72$		$\gamma = .13$		$\gamma = .24$		$\gamma = -.16$	
Females								
1	22	37	a	56	22	59	a	60
2–4	70	56	a	33	59	37	a	40
5+	7	7	a	11	19	4	a	0
	99	100		100	100	100		100
N	(27)	(268)	(4)	(18)	(54)	(430)	(3)	(58)
	$\gamma = .27$				$\gamma = .65$			

^{a}Insufficient cases for percentaging.

more, so we must control on age. Therefore the only way to test our hypothesis is to compare men who have fought or been arrested with nonfighters and non-arrestees who are the same age and have the same income.

In this table we are not primarily interested in proving that fighters or arrestees make less money; the median income of men who have been in a fight is only $5411 and of arrestees $5152, while those who have done neither have median incomes of $5864. What we want to show is that even compared to people with equally low incomes, aggressive people change jobs more often.

For example, among young men who earned less than $6000, 33% of those who have not been arrested have been on the same job for 5 years compared to only 6% of arrestees. Similarly, 40% of those who have been arrested have had five or more jobs while only 9% of those who have not been arrested have had this many jobs. The results are consistent for three of the four subcategories of men and both the subcategories of women where there are enough respondents to make an adequate test. (There are very few high-salaried women in the sample.)

Beneath each set of percentages there is a summary measure of association, which describes the strength of the relationship between being arrested and number of jobs held. The statistic is indicated by gamma (γ), a measure of association that is zero if the two questions have no relationship to each other and goes from $+1$ to -1 in the extreme case where a response on one of the questions is a necessary condition for a particular response to occur on the other questions. Gammas are in general larger than correlation coefficients; over the range of most of the correlations presented in this book, one can assume that gamma is approximately twice as large as the correlation coefficient that would be computed.

In the typical survey, measurement error is very large. Many of our respondents are completely inexperienced at answering these kinds of questions, and indeed it is extremely difficult, if not impossible, to word questions in such a way that every person asked them can understand what they mean. When we add to this the natural tendency of many respondents to try to give answers which they think will please the interviewer and then consider the sheer complexity of such concepts as aggression, we are not surprised that most of the associations reported here are small compared to perfection. Our very largest relationships have gammas between .3 and .5. (The gammas of .6 and .7 in Table 4.1 are the largest in this entire book.) Gammas on the order of .2 we treat as being fairly important and we will probably not read any gammas below .1 as representing a true association between two variables. In this sample size, a gamma of .12 would usually be significant at the 5% level with a one-tailed test. That is, if we predict in advance that there should be an association between two variables in a certain direction and that association appears and is over .12, there is less than one chance in twenty that the association would disappear if we took a very large sample rather than the 1700 unweighted cases that we have.

One might argue that being arrested frequently forces a job change, since a jail sentence or even a brief interruption to stand trial may cost a person his job. However it is very unlikely that the majority of people who have been arrested have been arrested three or four times. Thus most of the job changes shown in the table are not a result of arrest. This interpretation is supported by the finding that respondents who say they have been in a fight are also more likely to change jobs. When income and age are controlled, we again have six subcategories with sufficient cases for analysis, and all six of the gammas are positive. For men, the gammas range from +.11 to .34, and the two gammas for women are .36 and .60.

Changing jobs is highly correlated with being poor. For example, among older men, those who made over $6000 in the year before the study usually have had only one job during the last 5 years while those who earned less than $6000 are likely to have had more than one job. If the percentages in the table were given in the opposite direction, one would see that 95% of the men over age 30 who have had five or more jobs in the last 5 years earned less than $6000 a year, while 56% of those who had only one job in that period of time earned over $6000. We assume that low-paying jobs are more unstable, and hence people with few skills and low incomes change jobs more. It is also likely that frequent voluntary job-changing holds one's income down by decreasing chances for advancement and accumulating seniority. Thus we conclude that poor control of aggression causes job-changing which in turn causes (to some unknown degree) lower income.

It seems hostile people change jobs more often and lose income in the process. This does not mean that the respondents become angry at their bosses, tell them off, and get themselves fired. There is a much more subtle but equally effective way of expressing hostility: they simply do not show up for work.

The problem of tolerating frustration and controlling aggression becomes even more severe when a man's success depends upon being able to deal with white people. It is true that the good jobs are "white" ones. The 1960 census tabulates the racial composition of each of the 600-odd jobs in its detailed occupational listing. Each of our respondents was classified into one of these 600 categories, and the general pattern was that men who had been in fights or been arrested tended to work in jobs where many blacks were employed. The "pioneers" into mostly white jobs had not fought or been arrested.

Since better-educated blacks are in jobs which have fewer blacks in them, we have divided the sample into two groups: those who have either white-collar or better blue-collar jobs and those who have the poorer blue-collar jobs, in order to control on the respondent's job qualifications. Among workers with skilled blue-collar or white-collar jobs, the association between being arrested and being in an occupation where there are many blacks is. 19; for fighting, the association is .15. If we look only at semiskilled and unskilled blue-collar jobs, we find the same tendency: those blacks who work in occupations where there are few other blacks are less likely to have been arrested (γ = .15) and less likely to have been in a fight (γ = .05).

There seems to be three reasons why aggressive blacks do not hold "white" jobs. One is simply that an employer takes an arrest record more seriously when considering an applicant for a "white" job. Secondly, he may be more reluctant to hire the hostile black simply because of a certain surliness or some other aspect of his demeanor which indicates his hostility. Finally, it may also be that the man who has difficulty handling his hostility will be reluctant to apply for an opening. The threat of being placed in too close proximity to whites or being a "showcase black" may be too much for him to handle, and he may instinctively know it.

Inhibition of Aggression

Acting out of hostility is not the only serious problem facing our respondents; the opposite problem is equally unpleasant.

The traditional southern Negro—singing and "yassuhing" his way through a miserable life—is the classic example of inhibition of aggression. His inhibition might be due to a fear of being attacked for being "uppity;" it might be in response to his fear of a near-homicidal need to retaliate against whites. From a psychoanalytic viewpoint, he might be controlling anger aimed at a parent. From the viewpoint of a psychologist such as Leonard Berkowitz, he might have learned at the unconscious level to inhibit anger by imitating his parents, who were consciously "careful" in talking to or about whites.[9]

The argument is a difficult one to understand. Clinical psychologists (like Kardiner) talk about aggression as if it could be expressed in an almost infinite number of ways. The contention that people who have deep needs to express aggression are the ones who most inhibit their expression of aggression seems almost to defy a rational means of testing. However the idea that blacks are motivated by intense fears of white people, alternating with intense hostility toward them, may not seem so unreasonable when we have finished the analysis in this and the next few chapters.

For this part of the analysis we will use the response to a single question: "Did anything happen in the last month to make you angry?" (If a negative response is given: "Do you remember the last time something happened to make you angry?") Those respondents who answered "yes" to either of these questions were requested to describe what and who caused their anger.

[9]Leonard Berkowitz (1962), *Aggression: A Social Psychological Analysis,* McGraw-Hill, New York.

At first glance, the question seems to measure the needs of the respondent to express anger. Hostile personalities can be expected to have gotten angry recently. In fact, the question measures the opposite. It measures the need of inhibited people to conceal their anger from themselves and the interviewer. To begin with, almost everyone experiences in any one month enough frustration to become angry at least once. We also expect that he will remember the incident when asked about it. Of course even if a person was so fortunate as to have nothing frustrating happen in any important way in the last month, he should have no difficulty remembering the last time he experienced an emotion as strong as anger. If things were what they seemed to be, everyone should have been able to recall "the last time something happened to make you angry." In fact, however, one-third of our sample was not.

The anger question was developed and administered to both whites and blacks by Norman M. Bradburn.[10] Bradburn's study was not a single survey, but five separate surveys using the same questionnaire. The five separate surveys were taken in Prince George's County, Maryland, a white suburb near Washington, D.C.; a working-class area on the northwest side of Chicago; a random sample from the ten largest metropolitan areas of the United States; Warren, Michigan, which is a working-class white suburb of Detroit; and the Detroit ghetto. Bradburn's wording differs from ours; his specific wording asks only if the respondent recalls being angry "in the past few weeks." The data from Bradburn's study (given in Table 4.2) are clear. The respondents in Prince George's County and Warren, are the ones who are most likely to recall being angry. The blacks in the Detroit ghetto are considerably less likely to remember. It is obvious that life in Prince

TABLE 4.2

Recall of Anger by Various Samples in Psychological Well-Being Survey[a]

Sample	Recollection of anger (%)	
Prince George's County, Md.	59	($N = 1277$)
Ten largest metropolitan areas	54	($N = 270$)
Chicago	46	($N = 284$)
Warren, Mich.	61	($N = 542$)
Detroit	37	($N = 447$)

[a]Unpublished tabulations from Norman M. Bradburn's study. See *The Structure of Psychological Well-Being, op. cit.* for description of survey.

[10] Norman M. Bradburn and C. Edward Noll (1969), *The Structure of Psychological Well-Being,* p. 268, Aldine, Chicago.

George's County is not more frustrating than life in the Detroit ghetto, nor are the people in Prince George's County more hostile; yet we find that five respondents out of eight in the ghetto do not recall being angry, compared to only two out of five in Prince George's County.

Aggression may also be displaced from the true source of frustration to a safer target. The frustrated person may "kick the dog." Table 4.3 (also from Bradburn) suggests that blacks do displace aggression. The response was coded in terms of the target of the respondents' anger. We have divided these responses into two groups: those who become angry at their family, friends, or neighbors, and those who became angry at co-workers, employers, customers, or institutions or firms with which they dealt. For inner-city blacks, the first group is almost certain to be other blacks; in most cases the second group will be whites. Across the first four samples the predominantly white respondents were just as likely to have been angry at family, friends, or neighbors as they were to have become angry at the larger environment. But when we look at Detroit we see the black respondents are twice as likely to have been angry at family, friends, or neighbors. One might argue that the high level of anger directed toward family and friends reflects nothing more than the low level of interpersonal skill and the high degree of "social disability" which is often seen in the ghetto. However the important point is the very small amount of anger which is expressed toward co-workers, employers, and the government bureaucracies. Whites are three times as likely as blacks to become angry at work. In the survey only the whites said they were angry because of the way in which the larger world mistreated them or discriminated against them. This is also supported by the fact that only 6% of the blacks answered "yes" to the question, "In the past few weeks, were you treated badly by someone?" compared to 10% of the whites in Bradburn's survey.

TABLE 4.3
Direction of Anger in Various Samples of Psychological Well-Being Survey[a]

	Direction of Anger (%)			
Sample	Family or friends	Environment	No answer	Total (N)
Prince George's County, Md.	43	44	15	102 (767)
Ten largest metropolitan areas	41	54	14	109 (157)
Chicago	37	44	19	100 (131)
Warren, Mich.	47	43	13	103 (340)
Detroit	55	24	24	103 (170)

[a]*Ibid.*, unpublished tabulations.

At this point we have left the realm of traditional survey analysis and plunged into an effort to read personality structure from questionnaire responses. While we think we understand what is happening when a respondent says he does or does not remember being angry, there are some points of possible misunderstanding. For example, we are not certain that the word angry means the same thing in white Prince George's as it does in black Detroit. For that matter, it may be that the Prince George's office worker does have more things to get angry about at his or her job than the Detroit factory worker. We certainly hope that other researchers examine this question.

Who Gets Angry?

Let us return to our own survey data to see who it is who cannot recall being angry. At this point, the reader is perhaps anticipating the answer to this question will be the exact opposite of that dictated by common sense.[11] Table 4.4 seems to show that those people who are most subject to racism—men, the poorly educated, and Southerners—are the least angry. The table gives the percentage who do *not* recall being angry, either in the last month or earlier. The lowest numbers appear in the lower left part of the table and apply to northern-born women. This means that northern-born women are *most* likely to express and recall anger. Comparing the top two lines to the bottom two, we see that in every case, men are less likely to recall anger than women. Poorly educated men recall anger less, and southern-raised men recall anger less. Of the three factors,

TABLE 4.4
Repressed Anger by Age Moved North, Sex, Education

Sex and education	Those not recalling being angry (%)[a]		
	Born in North	Born in South, moved north before 10	Born in South, moved north at 10 or older
MALE			
Low education	34 (230)	43 (112)	48 (663)
High education	29 (421)	34 (134)	37 (325)
FEMALE			
Low education	22 (317)	24 (177)	38 (710)
High education	22 (363)	23 (141)	26 (553)

[a]Figures in parentheses are the numbers on which percentages are based.

[11]One is reminded of Stuart Chase's definition of common sense: "That sense which tells you that the world is flat."

education is the least important; sex and being raised in the North are about equal in their effects.

The sharp differences in Table 4.4 are probably understatements; the real differences in inhibition of aggression are probably even larger. The reason we say this is that the quality of the questionnaire item is poor.[12] It is a single item rather than a scale, and it is subject to a certain amount of fluctuation due to changing conditions. (Occasionally even the most inhibited person feels frustration severely enough to openly express his anger.) The item does, unfortunately, measure a bit of what common sense suggests: hostile people do get angry more. For example, people who have been in fights are somewhat more likely to recall being angry. However the effects of being from the North, or of being female, are more important than the correlation with being in a fight. All of this leads us to conclude that the anger item measures hostility or level of frustration to some degree, but mostly it reflects the inhibition of aggression.

The Consequences of Inhibition of Aggression

Inhibition of aggression is highly valued in most cultures and is certainly functional in an industrialized society. Inhibition of hostility is what makes it possible for a man to spend 40 years on the same job, or a woman to spend 40 years married to the same man. We admire the well-mannered child and a large part of what we are admiring is his ability to inhibit his aggressive tendencies. We think of inhibition of aggression as a middle-class or "feminine" virtue; this is why the extent of feminine anger in Table 4.4 is surprising. Nevertheless we are convinced that in the case of blacks, we are witnessing an excessive inhibition. As we shall see, the behavior and attitudes of the inhibited black are very much those of the stereotype of the happy slave. For example, respondents were asked a series of questions about the racial composition of the neighborhood they lived in and the kind of neighborhood they would prefer. One of the questions was, "Suppose someone came to you and told you that you could rent or buy a nice house, that you could afford, but it was in an all-white neighborhood and you might have some trouble out there. Are you the pioneering type who would move into a difficult situation like that?" Only 44% of the inhibited men said they would be pioneers, compared to 56% of the men who recalled being angry. The comparable percentages for women are 34 and 45%.[13]

---

[12]In general, an increase in error in a variable leads to a decrease in the correlation of that variable with other variables.

[13]Note that men are more willing to move into an all-white neighborhood than women. This may be simply because they are less afraid of physical harm; it may also be because they would be able to escape the pains of racial friction by going to work during the day or because they have less responsibility for protecting children.

Why are the inhibited blacks unwilling to be pioneers? There are several possible explanations. One explanation is that people who have difficulty expressing aggression must avoid putting themselves in aggression-provoking situations. This might be because bottling up their feelings of aggression creates inordinate psychic pain. A second possibility is that they avoid provocative situations for the same reason that they inhibit their anger: they are afraid of unleashing the demons that they have locked inside themselves, they imagine that others will attack them violently if they do express their anger. A third possible explanation is that continued inhibition of aggression has its own direct consequence—it leads to a sense of lethargy.[14] If we had the data, we could, for example, test the hypothesis that our inhibited respondents sleep more than others. We have no way to choose between these three alternative hypotheses. (There is a possible fourth explanation: blacks who inhibit their aggression do so because they are afraid of whites, and this is why they do not move into white neighborhoods. However, our data lend no support to this last hypothesis.)

All three of these specific interpretations lead to the same general hypothesis: that those who cannot recall being angry will avoid all forms of risky or potentially painful situations. Since a large fraction of the experiences one has in a society like this carry at least a slight risk of pain, this is an extremely powerful inhibitor of action. Table 4.5 provides three bits of data which support this hypothesis. First, it shows that the attitude item we have just looked at is reflected in actual behavior. The inhibited respondents are less likely to live in integrated neighborhoods. For example, in the second column, 30% of male high school graduates who can recall being angry live in integrated neighborhoods, compared to 23% of the inhibited males. (The differences, in fact, are not as large as they are in the case of the attitude, but this is not surprising. For one thing, at least some of our respondents are living in neighborhoods which were selected by their spouses.) The second line shows quite sharp differences in participation in civil rights demonstrations. This is exactly the kind of risky situation which fits the hypothesis well. The third line is very surprising as well as provocative. The respondent was asked, "Can you think of a particular company or employer where you are pretty sure you could get a job right now?" He was then asked to give the actual name of the employer, and only those respondents who could give an actual name are considered as having useful knowledge of another job.

Why are inhibited respondents less likely to be able to name potential employers? The answer supports the idea that inhibited people are more lethargic, less alive to the world around them. But it also fits the idea that they are avoiding anger-producing situations. After all, the act of applying for a job is certainly

---

[14]This is taken from a conversation with Robert A. Gordon, Department of Social Relations, The Johns Hopkins Univ., Baltimore.

TABLE 4.5
Correlates of Expression of Anger by Sex and Education

| | Sex, Education | | | |
| | Men | | Women | |
Recollection of Anger	High school graduate	Not high school graduate	High school graduate	Not high school graduate
Living in half white or mostly white neighborhood (%)				
Recall anger	31 (545)	30 (557)	32 (798)	28 (815)
Do not recall	29 (286)	23 (444)	26 (251)	24 (383)
Participating in civil rights demonstrations (%)				
Recall anger	29 (595)	13 (537)	19 (798)	7 (815)
Do not recall	19 (286)	9 (444)	13 (257)	7 (383)
Naming potential employer (%)				
Recall anger	27 (595)	31 (557)	37 (798)	20 (815)
Do not recall	29 (286)	17 (444)	27 (251)	14 (383)

risky. The easiest way to escape even thinking about it is to simply say, "No, I'm not interested in thinking about where there might be another job." To think about where there might be another job often means thinking about where there might *not* be another job. But perhaps best of all, these data suggest that inhibition of anger means inhibition of all forms of aggression. Any form of competition, of asserting oneself over others, is an aggressive act. The person who cannot express anger cannot compete with others.

However we cannot conclude that expression of anger is an indication of good overall adjustment. It has some advantages, but we must withold judgment until we have analyzed other personality characteristics.

The Two Faces of Mishandled Aggression

There is a chain gang song which says "seems like everything I do is wrong," and at this point it does sound as if we are making that kind of blanket accusation against northern blacks. On the one hand, we are saying that they express too much aggression and get into trouble; on the other hand, we are saying that they express too little aggression and are unable to compete. Does this "damned if you do and damned if you don't" criticism make any sense? We think it does. What we are arguing is that every black sees himself as a victim of white people's prejudice, discrimination, and violence. This is in many ways the most important thing in his life. White people control virtually all of the resources of the society. If a black wants something, whether it be a raise, a welfare check, or a house, he must either negotiate a bargain with the white man who controls it or, if it is something he has a right to such as the vote, he must trust white people to act fairly of their own volition. Given his experience and his feelings about whites, he might expect to be victimized in all these situations. The result is that he carries with him an enormous amount of hostility. The hostility becomes a source of constant tension for him. What we are saying is that a large number of blacks attempt to handle that tension by suppressing their aggressive tendencies. Another large group of blacks find that this internal hostility explodes, driving them to violence and trouble with the law. In fact, as we shall see, there are a significant number of blacks who have both of these problems at the same time. They are too passive most of the time, but occasionally explode and get themselves into trouble.

So far in this chapter we have used three questions from the questionnaire to measure "mishandling of aggression." There is a fourth one which we want to add at this point. The question is taken from a survey done in the late 1950's by Donald J. Bogue and the question simply asks, "When you hear about a Negro being discriminated against, does it bother you? When it does bother you, do you feel angry, or do you usually feel sad and depressed about it?"[15] The reader will see that this is another way of asking whether the respondent can express anger. If he can, he will become angry at discrimination. If he cannot, he will become sad or depressed as he internalizes this aggression. In response to this question, 49% of our respondents said that they are usually angry and 48% say that they are usually sad or depressed. The remaining 3% said they do not react in either direction.[16]

Being inhibited is not a guarantee against expressing aggression in unhealthy ways. In fact, the inhibited respondents are more likely to have gotten into a fight than the respondents who remember being angry. Following what we have already said, it is clear that the successful aggression-handler must be able to express anger. Certainly he should be able to recall easily the last time that he was angry. Let us add to that a second requirement: he must be able to be angry rather than depressed when he hears about an incident of discrimination. We think that internalizing aggression in this case is unfortunate for much the same reason that being unable to recall anger is unfortunate; it means that the respondent is too inhibited to assert himself. On the other hand, we would insist that

[15] Our version of the question was worded in two sentences to permit the respondent who at first says he has no feelings about it to still choose between anger and sadness. (Source: unpublished manuscripts of Donald J. Bogue, University of Chicago).

[16] The angry versus sad question correlates modestly with the recollection of anger questions: $\gamma = .25$ for men, $.29$ for women.

the successful aggression-handler should be able to live his adult life without ever being in a fight or being arrested.

If we use these criteria as our definition, how many successful aggression-handlers do we find in the sample? The rather depressing answer is given in Table 4.6. In that table we cross-classified fighting and being arrested with our two measures of inhibition of aggression to produce four categories: the *controlled* are successful aggression-handlers; they can both recall anger and become angry when they are bothered by discrimination, but they do not report being in a fight or being arrested.

The *uncontrolled* are those who do not inhibit aggression, but have either been in a fight or been arrested. The *inhibited* either cannot express anger about discrimination or cannot recall being angry, and have not fought or been arrested. The *explosive* are also inhibited, but have either fought or been arrested. Only 11% of the men and 18% of the women fall into the category of the controlled.

Of course these are not sacred numbers—a seemingly insignificant change in the wording of the questions might produce a considerable change in the percentages. Nor are we pretending that these four questions have the accuracy of a clinical test (which still would have major problems of interpretation). We do not have much sympathy with the now infamous studies which have used survey-style techniques to define most of the population as "sick" and only a fraction as "healthy." But it is nevertheless true that we cannot imagine any change in wording which would make the number of successful aggression-handlers more than a minority.

Furthermore, although we cannot apply precisely the angry-sad question to whites since it has to do with racial discrimination, we nevertheless can extrapolate on the basis of national surveys using the other three questions to see how whites would look under this same test. Bradburn has shown that only two-thirds to one-half as many whites inhibit their anger; our data show that only

TABLE 4.6

Distribution of Sample into Typology of Aggression-Handling

	Men (%)		Women (%)	
	Have neither fought nor been arrested	Either or both	Have neither fought nor been arrested	Either or both
Can recall anger, react with anger to discrimination	Controlled 11	Uncontrolled 15	Controlled 18	Uncontrolled 6
Only one or neither of above	Inhibited 36	Explosive 38	Inhibited 57	Explosive 19

one-third as many whites are involved in fights or have been arrested. Putting this together suggests that at least a third of whites would fall into the "controlled" aggression category. Obviously, one of the costs of being black is that handling one's aggression in a healthy fashion is much more difficult. This is the only safe conclusion we can draw from Table 4.6, but it is an important one.

Aggression and Racial Attitudes

Psychoanalytic theory argues that we should think of the fighter as someone who directs his anger toward an external object, while the person who does not consciously get angry directs his anger inward. And Kardiner and others argue that blacks misdirect aggression in order to avoid expressing aggression towards whites. We can now test this notion by examining the relationship between aggression-handling and feelings about whites and blacks. Six questions were combined into a scale measuring anti-white feeling.[17] Two other items, both of which measure the belief that Negroes are lazy, make up an anti-black attitude scale.[18]

Table 4.7 gives the data, and the results turn out more or less as predicted. Inhibited men show the least anti-white sentiment and the most anti-black feelings. The controlled show much more anti-white sentiment and much less anti-black feeling. Among those who have fought or been arrested, those who cannot recall being angry—the exploders—are less anti-white and more anti-black than those who can recall anger.

For women, the pattern is somewhat different. Inhibited women, like inhibited men, show little anti-white feeling and high amounts of anti-black feeling, and controlled women are less anti-black and more anti-white. The point of divergence between the sexes lies in the behavior of women who have fought or been arrested and recall being angry—the uncontrolled group. A high percentage of these are anti-black; perhaps they are also externalizing by blaming their troubles on black men.

---

[17]The six items are (answering "frequently" or "sometimes" to "when I am around white people...."): (1) "I am afraid I might tell him what I really think about white people;" (2) "I am afraid I might lose my temper at something he says;" and (agreeing to): (3) "Sometimes I'd like to get even with white people for all they have done to the Negro;" (4) "The trouble with white people is they think they are better than other people;" (5) "If a Negro is wise, he will think twice before he trusts a white man as he would another Negro;" (6) "There are very few, if any, white people who are really unprejudiced" (see Appendix 2 for intercorrelations).

[18]The agree-disagree items are: "generally speaking, a lot of Negroes are lazy;" and "a lot of Negroes blame white people for their position in life, but the average Negro doesn't work hard enough in school and in his job."

TABLE 4.7
*Aggression Handling and Anti-white and Anti-black
Attitudes by Sex*

Aggression type	High anti-white feelings (%)	High anti-black feelings (%)	N
MEN			
Inhibited	21	30	(684)
Controlled	36	20	(204)
Explosive	28	29	(707)
Uncontrolled	40	23	(291)
WOMEN			
Inhibited	19	36	(1294)
Controlled	25	28	(410)
Explosive	25	32	(429)
Uncontrolled	25	40	(134)

The Childhood Origins of Aggression-Handling

From our everyday experience, we would guess that patterns of aggressive behavior reflect relatively stable personality traits. We do not expect a person who is mild-mannered one year to be bad-tempered the next.

The stability of aggression-handling is reflected in its association with characteristics of the respondent's parents. For example the man whose mother attended high school is twice as likely to fall into the "controlled" category as the man whose mother did not go to high school (the percentages are 15 and 7%, respectively).

This is not simply because men with better-educated mothers are themselves better educated; the mother's education is a better indicator of controlled aggression-handling than the amount of his own schooling.

We saw in Chapter 2 that northern-born respondents were more likely to fight or be arrested, and in the first part of this chapter we saw that Northerners were more likely to express anger. Putting these two findings together with the data on the mother's and the respondents education, we can draw portraits of the "ideal type" man or woman in each category: The *inhibited* respondent is an older Southerner, a high school graduate whose parents had little schooling. The *controlled* person is young, northern-born, a high school graduate whose parents attended high school. The *explosive* respondent is a southern high school dropout with poorly educated parents. The *uncontrolled* is a young Northerner, a high school graduate with poorly educated parents.

Summary

The handling of aggression is a major personality problem for blacks since racial discrimination, which is itself an expression of hostility, causes an identity crisis in its victims. Having been arrested or in a fight are indicators of aggressive behavior and the data have shown that they are related to job-changing, low income, and "traditional" employment. The inhibition of aggression is also a problem, since the inability to express anger may imply a denial of reality. A typology of aggressive handling reveals four categories of adaptations among blacks. The types are related to black racial attitudes and reactions to discrimination.

This analysis is one of the first which has attempted to analyze inhibition of aggression using "hard" methodology. We hope it will encourage others to do further work.

5

Internal Control

There was a difference between the Northern Negroes and me. It wasn't that they were more intelligent. More sophisticated, I guess that's what you'd call it—they expected to be able to learn, and I didn't . . .[1]

Southern black who moved North to attend college

The next task is to examine a variable we will call "internal control of environment," as measured by a scale of five questions. Table 5.1 gives the wording of the five questions and the percentage of blacks and whites who gave the answer indicating "high" internal control to each question.[2] The intercorrelations between the five responses are given in Appendix 2. The questions come from two different sources: the first three are from *Equality of Educational Opportunity,* usually called the Coleman Report; the last two are from a scale developed by Julian P. Rotter.[3] The five items all deal with the general question of how a

[1] Bertram P. Karon (1958), *The Negro Personality,* p. 2, Springer, New York.

[2] James S. Coleman, Ernest Q. Campbell, Carol J. Hobson, James McPartland, Alexander M. Mood, Frederic D. Weinfeld, and Robert L. York (1966), *Equality of Educational Opportunity,* p. 288, U.S. Office of Education, Washington, D.C. The words, "Good luck is as important as hard work for success" are changed from Coleman who used "more important than"; the change was made to increase the number agreeing with the statement.

[3] Julian B. Rotter (1966), "Generalized Expectancies for Internal Versus External Control of Reinforcement," *Psych. Monographs,* **80,** 609. In all, four Rotter items were administered to blacks and two of these to whites. There was essentially no difference in the responses of whites and blacks on either of the two Rotter items which were given to both

TABLE 5.1
The Internal Control Scale

Items (high control response in parentheses)	High internal control response (%)	
	Blacks (N = 3968)	Whites (N = 1288)
1. Good luck is just as important as hard work for success. (Disagree)	40	55
2. Very often, when I try to get ahead, something or somebody stops me. (Disagree)	43	63
3. People like me don't have a very good chance to be really successful in life. (Disagree)	69	78
4. Being a success is mainly a matter of hard work, and luck has little or nothing to do with it or—getting a good job depends mainly upon being in the right place at the right time.	65	54
5. When I make plans, I am almost certain that I can make them work or—it is not always wise to plan too far ahead because many things turn out to be a matter of good or bad fortune anyhow.	50	*

*Item not asked of whites.

---

groups. Two of the four items were dropped because they correlated poorly with the others, including one of the two given to whites. Apparently the intercorrelations depend on what population is used. The four items were selected from a larger scale, based on the fact that they correlated best with the total scale when administered to white college students. Patricia Gurin administered the total scale to black college students and factor analyzed it; one of the two items we kept and one we rejected fell into her first factor and the other two items fell into a second factor—exactly opposite of our result. (See Patricia Gurin (1969), "Internal-External Control in the Motivational Dynamics of Negro Youth," *J. Soc. Issues, XXV* (No. 3), 29-53. The two items which we dropped from our study are: "Many times I feel that I have little influence over the things that happen to me . . . It is impossible for me to believe that chance or luck plays an important role in my life;" and "Most people don't realize the extent to which their lives are controlled by accidental happenings . . . or, there really is no such thing as 'luck.'" Perhaps the reason is that neither item refers to occupational or other kinds of success, and people may entertain superstitions regarding luck in cards or love without letting it influence their career planning.

person feels about his power to control his life—will planning and working pay off, or is the future really subject to the whims of fate?

Not surprisingly, blacks are noticeably more fatalistic than whites on the first three items. One might argue that blacks give more fatalistic answers because they are realistically more pessimistic about their life chances, but the first two items do not easily admit to this sort of interpretation.

There are four questions we must now attempt to answer: (1) What is the theoretical rationale for this scale? What is it supposed to measure? (2) Is it associated with the respondent's other attitudes and his behavior in the way that we expect? (3) Can we interpret these associations as meaning internal control causes the behavior, or is there some other explanation for the relationship? (4) Can we infer from these correlations what the personality differences are between the people who have this attitude and those who do not?

The Concept

Rotter calls the choice between the two alternatives in the scale a choice between internal and external control of the environment. If an individual says that his plans may succeed, then he believes he has control over that part of the environment he is trying to master; if he says that good or bad luck will decide his fate, then he believes his environment is controlled by external forces (i.e., he has "low" internal control). At issue is whether the subject engages in what Rotter calls "social learning." Life is from this point of view a sort of Skinnerian teaching machine, where the child puts a nickel in his piggybank, reopens it later to find the nickel still there, and decides that if he saves a nickel, he can spend it later. But all of this implies a sort of basic belief that he can act (put the nickel in the bank) and thereby cause the desired effect (in this case, the nickel reappearing when he wants it). Suppose that he does not believe it is possible for him to create such an effect; then he will not attempt to learn—in this case to save money. If he believes that jobs are a matter of luck, he will not learn to apply for them.

If this is the right interpretation, then subjects who do not have internal control will not be able to learn to plan for the future; they will not defer gratification. Table 5.2 gives strong support for this contention. This table shows that people with internal control are considerably more likely to have a checking account. We have presented data only for married and widowed respondents who earned over $3500 the previous year. We have divided respondents by sex and whether their family received over or under $8500 income. Thus there are four different comparisons we can make in the table, and all four show the predicted association.

The high internal control response is associated not only with this measure of saving money, but with a number or related items as well (Table 5.3). In Table

TABLE 5.2
Internal Control and Having a Checking Account, by Sex and Income[a]

Internal Control	Percent of married or widowed respondents with following incomes having checking accounts	
	$3500-8500/year	$8500/year or more
MEN		
Low control	16% (140)	53% (58)
Medium	36% (331)	60% (220)
High	44% (214)	75% (174)
WOMEN		
Low control	29% (194)	44% (45)
Medium	35% (417)	58% (154)
High	45% (259)	78% (155)

[a]Married and widowed respondents only. Single, separated, and divorced respondents dropped because of inadequate number of cases.

5.2 we controlled on income to show that the large number of checking accounts held by high internal control subjects was not merely the result of their higher income. Actually the respondent's education predicts his money-saving habits better than his income. In table 5.3, therefore, we controlled on education as well as sex, and where necessary, marital status, age, or employment status.

In the first row of Table 5.3 we see that married respondents with high internal control are more likely to own their own home. For example, only 19% of the males with low internal control who did not finish high school own homes, compared to 24% of the medium-control and 38% of the high-control subjects. The overall association is a gamma of .29 summarizing the relationship for poorly educated men. Looking across the first row, we see that the relationship between internal control and home-buying is consistent for all four groups, although it is weaker for women than for men. If we succeed in showing that internal control is indeed a cause of home-buying, then we can also draw the conclusion that the attitudes of the husband are more important than those of the wife in determining whether the couple will buy a home or not.

The second row is the association between control and holding less than three jobs in the past 5 years. Recall that in the preceding chapter we argued that job-changing was one of the factors which reduced the incomes of aggressive men.

TABLE 5.3
Internal Control and Six Types of Behavior

	Association (γ) between internal control and six measures of behavior.			
	Men		Women	
	Nongraduate	High school graduate	Nongraduate	High school graduate
Owns home[a]	.29	.22	.12	.19
Held less than 3 jobs in past 5 years[b]	.13	.01	.39	.14
Scores 2+ on finance scale	.26	.33	.27	.35
Can pay off consumer debt	.08	.24	.16	.12
Never evicted	.28	.54	.30	.14
Never had utilities cut off	.03	.33	.07	.02

[a]Married and widowed respondents only.
[b]Employed respondents over 30 only.

Here we see that both men and women with high internal control change jobs less, although the differences are small for men. Since young people change jobs often in settling into the labor market, these data are only for employed men and women over 30.

We asked the respondent if he or she had life insurance, a savings account, or owned savings bonds or stocks. Together with the checking account question, this yielded a 5-item scale of financial behavior. In general, large numbers of subjects seem *not* to have savings; over a third had neither a checking nor a savings account. In row 3, we correlate control with having two or more of these five items, on the assumption that adequate planning would require that the subject have both life insurance and some sort of bank account, at the minimum. As we would expect, the data show large associations for both sexes. High school dropouts with high internal control have saving habits as strong as those of high school graduates with low control.

The respondents were asked the amount of their monthly credit payments, exclusive of their house mortgage, and whether they could now pay off their debts from their savings. Most blacks are "in the red"—only a third could pay off their debts instantly. Education makes little difference here; the high school graduates are no more likely to be able to pay off their debts than those without diplomas. But internal control *does* make a difference.

The last two rows use the responses to two questions: "Have you ever had to move from a house or apartment because you couldn't pay the rent or mortgage?" and "Have you ever had your gas, lights, or water turned off?"

In summary, all 24 measures of association are in the expected direction, and ten of them are over .25. Apparently internal control is linked to deferring gratification, planning and saving, as predicted by the theory. Later we shall try to demonstrate that this attitude of internal control is a *cause* of these habits.

Alienation and Rebellion

Melvin Seeman worked with Rotter in developing the concept and measure of internal control but has put forth a somewhat different interpretation, by referring to it as a form of alienation which he calls powerlessness.[4] By calling it alienation, he makes us realize that the sense of futility, of effect not following cause, can be seen as a root of the rebelliousness of black youth. Coleman comments to this effect in a response to one of Seeman's articles.[5] Seeman had just shown that the internal control scale, administered to reformatory inmates, could predict which inmates would learn the most about parole procedures. Earlier he had shown that high-control tuberculosis patients learned the most about the treatment and cure of the disease.[6] As Coleman puts it, "Seeman's results suggest a general phenomenon: that a man is sensitive to the cues of his environment only when he believes he can have some effect on it. He will learn only when such learning will benefit him." This is close to Rotter's concept.

Coleman goes on to argue that the traditional southern plantation economy gave blacks very little opportunity to control any aspect of their environment. "The plantation Negro's condition may not have been good, but it was certain, and he need have no concern for his future, for he could not affect it."[7] From this starting point, Coleman then went on to use the variable in *Equality of Educational Opportunity.* (For the reader with an interest in the workings of science, I will note that we added it to our questionnaire at Coleman's suggestion.) Bonnie Bullough, a student of Seeman's, devised a piece of research also building on Coleman's idea.[8] She argued that if internal control enables one to learn about parole or cures for tuberculosis, it could also serve to enable blacks to escape from the ghetto. She compared middle-class blacks who lived in the ghetto with those living in predominantly white areas and showed that the integrated blacks had higher internal control. Furthermore, she showed for both groups that living in an integrated neighborhood as a child was associated with a high sense of control.

[4]Melvin Seeman (Dec. 1959), "On the Meaning of Alienation," *Amer. Soc. Rev. XXIV,* 783-791.

[5]James S. Coleman (July 1964), "Seeman's 'Alienation and Social Learning in a Reformatory', Two Reactions," *Amer. J. Soc., LXXX* (No. 1), 76-78.

[6]Melvin Seeman and J. W. Evans (Dec. 1962), "Alienation and Social Learning in a Hospital Setting," *Amer. Soc. Rev. XXVII,* 772-782.

[7]Coleman, *op. cit.,* p. 77.

[8]Bonnie Bullough (March 1967), "Alienation in the Ghetto," *Amer. J. Soc.,* 72, 469-478.

It seems to us an easy step to move from this conception of internal control back to traditional alienation theory. If having high internal control means that one believes one can, by trying, perhaps escape the tuberculosis ward, the reformatory, or the ghetto, then having low control means that one does not believe he can escape. Living is a passive exercise in which one receives good or bad luck. And the luck is mostly bad if one believes that "everytime I try to get ahead, something or somebody stops me."

It seems logical that the appropriate response is to give up—to live for the present, "for tomorrow we die." The ghetto's "street culture" is full of precisely this sort of simple hedonism. Why go to school? It won't help you later. Why work hard? The boss may fire you anyway. Why do anything, unless it feels good now? This hedonism is reflected even in the language—phrases like the greeting "How're you feeling?" To be "cool" means to be uninvolved, uncommitted, immune to disappointment. From this point of view, night-clubbing, liquor, the numbers, what in general was once called "kicks," can be seen as attempts at instantaneous gratification in an uncontrollable environment.[9]

Logically it is an easy step from this to the hypothesis that low control of environment carries with it an urge toward irresponsibility and rebelliousness. This interpretation is supported by Table 5.4 which shows that both men and women with low internal control are more likely to be in fights or be arrested.

While up to now we have interpreted arrest as being an indication of acting out aggression, it also seems likely that many arrests derive from drunkenness, and hence an arrest record may also be associated with escapist behavior through alcohol (or drugs). Logically, low internal control should be associated with escapism, but we have no data on this.

TABLE 5.4
Internal Control by Fight and Arrest

| | Association (γ) between internal control and fighting and being arrested | | | |
| | Men | | Women | |
	Nongraduate	High school graduate	Nongraduate	High school graduate
Never in fight	.05	.18	.13	.09
Never arrested	.07	.37	.15	−.07

[9] See Harold Finestone, (1957), "Cats, Kicks, and Color," *Social Problems,* **5,** 3-13. For a valuable portrait of the street culture, see Malcolm X (1964), *The Autobiography of Malcolm X,* Grove Press, New York.

Dealing with Authority, "The Man," and "Passing the Buck"

One other theoretical orientation enters here in a very similar scale called "personal control," developed by Fred L. Strodtbeck.[10] He was concerned with the differences in the achievement of various ethnic groups, and hypothesized that differences in socialization patterns might be the answer. He developed a technique, "revealed differences," in which he administered a questionnaire to a family triangle—mother, father, and son—and showed them the items they disagree on and asked them to discuss their differences. He taped the discussion and coded it to measure the relative power of the three family members. He found that Jewish boys have more influence in these discussions than do the boys in a matched sample of Italian-American families. He then showed that boys in each ethnic group who were dominated by their parents in the discussion have a high sense of being dominated by fate. The reasoning is simple: the child learns that his father or mother is powerful, domineering, and sometimes capricious; he then transfers this perception of his parents to his dealing with other authorities, and assumes that his life will always be dominated by the invisible hand of authority —fate, luck, or God.

This is a simple and rather persuasive notion, which seems very applicable to blacks. First, the child learns by observing his parents that a powerful figure lies outside of the family: "the man" is a sort of invisible superfather from whom all blessings flow, if and when he chooses to dispense them. Second, we know that at least in the South, black parents are strict in teaching their children racial etiquette, which may simply mean teaching the child to be quiet, timid, cautious, and deferential. This means the child sees himself as playing a passive role *vis à vis* his parents.

All of this means that learning that whites are dominant and blacks weak takes the form of learning that one's life is controlled by authority figures—parents, teachers, policemen, bosses—who are all-powerful. Hence the person with low internal control may have great difficulty in handling his relationships with authority. Presumably this means that the black with low control would become anxious and angry in his dealings with authority. We have no direct way to test this hypothesis with our data. It is true that persons with low internal control are more likely to say they are dissatisfied with their supervision at work, but this is rather weak evidence, for people with low control complain about other aspects of their job as well, although not as strongly. (Internal control correlated .21 for men and .24 for women with a job satisfaction scale constructed from all these items.)

[10] Fred L. Strodtbeck (1958), "Family Interaction, Values, and Achievement" in D. C. McClelland (ed.) *Talent and Society,* pp. 135-194, Van Nostrand, Princeton.

Persons with low internal control are also more likely to complain of mistreatment by police. Respondents were given a list of types of people, for example, police, teachers, landlords, social workers, etc. They were asked if they had had contact with whites in these roles lately, and asked if they felt they were treated differently because they were black. The percentage who said "yes," that the white had treated them differently than he would have a white person in their position, ranged from a low of 12% for social worker and landlord to a high of 18% for policeman. The results are shown in Table 5.5.

But of particular interest to us is that persons with low internal control are considerably more likely to perceive themselves as being mistreated by the police. Thirty-one percent of the low-control men perceive mistreatment, compared to only 21% of men with high control. The differences for women are even sharper: 15 versus 6%. Of course, part of this difference may be "real," in that low-control people are poorer and less well-educated, and presumably the police are harsher in dealing with the lower classes. But the differences are too large to be passed over in this way.

Despite our inability to test Strodtbeck's argument with our data, it does seem a plausible way to connect the caste status of being black with low internal control. The person who sees himself at the mercy of the white man is making a simple generalization to see his life as controlled by capricious forces in general. Or it can be argued in the opposite way: the black person who, for any reason, sees himself as unable to control his future, who opts for short sighted hedonism in defense, can blame his troubles on whites.

Table 5.6 shows that people with low internal control score very high in anti-white feeling. While only 4% of men with high control score high in anti-white

TABLE 5.5
Contact and Perceived Mistreatment by Whites

Roles played by whites	Those having recent contact with whites in this role (%)	Of those in contact, percentage who felt they were treated differently because they were black.
Policeman	37	18
Boss or supervisor	57	14
Co-worker	50	15
Social worker, or other government employee	28	12
School teacher	31	13
Store clerk or storekeeper	68	17
Landlord	25	12
Person on the street	47	15

TABLE 5.6
Internal Control and Anti-white Sentiment

Anti-white sentiment	Internal control (%)		
	Low (score 0-1)	Medium (2-3)	High (4-5)
MEN			
High (score 5-6)	17	11	4
Medium (3-4)	60	47	33
Low (0-2)	22	43	64
	99	101	101
N	(365)	(861)	(624)
			$\gamma = -.46$
WOMEN			
High	12	6	3
Medium	48	39	37
Low	40	55	60
	100	100	100
N	(546)	(1028)	(647)
			$\gamma = -.25$

feeling, 17% (four times as many) score high in the low control group. The association for men is a gamma of $-.46$, and this is the largest nontrivial association in our entire analysis.

One way to describe the magnitude of the association is to say that if it were larger, we would assume there was a measurement error biasing the results.[11] Another way to describe a gamma of $-.46$ is to say that if the items were not obviously measuring different things, we would be tempted to say that there is only one concept here. The items in the internal control scale are correlated as well with anti-white feeling items as they are with each other!

So we conclude that hostility toward whites and the feeling that whites are prejudiced is virtually the same thing as believing that life is controlled by luck and that making plans is futile. This seems to be conclusive evidence for our general thesis that relations between whites and blacks are in some way responsible for at least one attitude which is important for achievement. However we

[11] There is the possibility that since many of the items in both scales are scored low if the respondent agrees with them, the high association is due to "an acquiescent response set"—in simple English, the tendency some respondents have to agree to anything. However, if this were the case, the best-educated respondents, who are less gullible, should show lower correlations. In fact, the association between antiwhite feelings and internal control is a gamma of $-.55$ for high school graduate men, higher than the association for high school dropouts.

would like to know a good deal more about this relationship. In what way does the sense of being discriminated against tie into a fatalistic, hedonistic, and rebellious way of life?

The first question is, why is the association so much stronger for men than for women? The answer turns out to be rather simple. As many writers have pointed out, the black woman can achieve many of her life goals through marriage and child-rearing, within the racially segregated world of the family. It is the man who must go forth into "the white man's world." It is the male's life-chances which are tied to the race issue. Our data support this argument, for if it is right, then working women should be more like men than housewives are, and this is indeed the case. For women who do not work, the association between antiwhite feelings and sense of internal control is only −.22; for women who do work, the association is −.46, based on 748 cases (weighted, as always).

When we first saw the association between internal control and anti-white feeling, we assumed that a rather simple model would hold: blacks subject to the most discrimination would be most antiwhite, and this would in turn lower their self-esteem and sense of control over their environment. This is a plausible argument that is based on a simple direct link between "being treated as an inferior" and anti-white feeling. Unfortunately the world is not so simple, and the data do not support this hypothesis. For example, the argument assumes that blacks who have been subject to the most discrimination—those who attended segregated schools, for example—would be much more anti-white, but they are not. Apparently feelings about white people are derived in a more complex way. Untangling this riddle is a problem we shall return to in Chapter 10. So far we have shown one fact: a low sense of internal control is associated with a tendency to see authority figures, and the environment in general, as hostile. The low control subject in effect blames the world for his troubles and criticizes the way in which his supervisor, the police, and whites in general treat him.

Internal Control as a Cause of Behavior

Thus far we have shown that low internal control is associated with being unable to defer gratification, control aggression or avoid arrest, and being hostile toward authority figures. The next problem is to find out whether this attitude is a cause of these factors or merely an innocent correlate. There are two alternative hypotheses to consider:

1. Although internal control is associated with these kinds of behavior, there is some other factor involved which also causes internal control to be low. For example, it might be that poorly educated respondents behave in the fashion we have described and also have low internal control, but that education is the crucial factor. It is true that education is highly associated with internal control, as Table 5.7 shows; only one-fourth of the persons with low control are high school

TABLE 5.7
Internal Control by Education

Internal control	High school graduates (%)	
MALES		
Low	25	(365)
Medium	46	(875)
High	60	(626)
		$\gamma = .38$
FEMALES		
Low	29	(552)
Medium	42	(1036)
High	70	(651)
		$\gamma = .46$

graduates, compared to over half of the high-control subjects. But, as we have seen, education was controlled in the tables in this chapter.

Another possible factor is intelligence. The academic achievement of respondents was measured by a shortened version of a multiple choice synonym test developed by J. B. Minor.[12] "Verbal achievement," as we shall call what this test measures, is highly associated with control ($\gamma = .45$ for men and .44 for women). However it is not true that high control people behave the way they de because they are more intelligent. Persons who score high on the verbal achievement test do save their money, are evicted and arrested less, etc. But, in general, internal control is a better predictor of behavior than intelligence. Intelligence is not the answer either.

Of course, it is not possible to eliminate every conceivable possible factor which might be controlled. (It is for this reason that methodologists do not speak of "proving" a hypothesis, since it is logically impossible to do so.) But we have searched the rest of the questionnaire, and no other factors are as important as education and intelligence. It seems unlikely that internal control is a spurious factor.

2. The second alternative hypothesis is that a sense of high internal control is the *result* of the behavior, rather than the cause. The man with a bad job and a police record says he has had bad luck; the man with no record and with money in his pocket claims he earned it all through hard work.

---

[12] John B. Minor (1957), *Intelligence in the United States,* Springer, New York: The 20-item test was developed by R. L. Thorndike and used by both Thorndike and Minor with survey populations. The 20-item version of the test is itself a cutdown version of the vocabulary section of the Institute for Educational Research Intelligence test, CAVD. Thorndike has computed reliability coefficients ranging from .80 to .915 (see Minor, pp. 48-50). Our test used 9 items, one of which failed to correlate and was dropped. For details, see Appendix 3.

This alternative hypothesis does not seem to fit our data, however. First of all, low internal control is not strongly associated with those misfortunes over which the respondent has little control. For example, the association (without the education control) between internal control and being evicted is .41 for men; for being arrested, .25; having utilities cut off, .18; for being in a fight, .14; but it is only .11 with having gone hungry, and only .06 with being robbed. Similarity, of four good things in the respondent's life, internal control is more strongly associated with receiving a bonus or prize (.22) or "being praised for a good piece of work" (.18)—two things the respondent can perhaps achieve through his own efforts—than with "having enough money to do what you wanted" (.13) or having one's children do something to make one proud (.12). Also, if the hypothesis were true, internal control should be highly correlated with income, but that correlation is only moderate (.15). The association is much better with education, which is not a recent achievement. This seems in itself convincing evidence because if internal control went up or down in life, depending on what happened, it could not remain highly associated with a factor like education, which was fixed in the respondent's late teens. The argument is even stronger for intelligence. The high association with the verbal achievement score, which is presumably relatively constant from childhood through adulthood, suggests that intelligence and a sense of internal control both were established in childhood, with relatively little change since. It seems impossible for internal control to change very much and remain highly correlated with intelligence for adults.

Thus it seems very likely that the sense of control over one's environment is an important building block in the personality, with far-reaching consequences. The next question is, what kinds of people have this sense of internal control?

Causes of Internal Control

We have already seen one important cause (or effect) of internal control: education. It seems likely that it is *both* cause and effect. School is where one learns to look at the world objectively, to analyze, to put aside superstition, to consider cause and effect. Internal control, as we have measured it, is quite simply the belief that one can cause the effect he wants to achieve. Hence, it makes sense that the more schooling one has, the more likely one is to score high on the internal control scale. But we can also argue that the kind of attitude that makes one save money and stay out of trouble is precisely the attitude that makes one stay in school. The difference between a dropout and a graduate is usually only one or two years of school. What does one learn in that one year to make such a big difference?

In Appendix 3, we have attempted to unravel this puzzle, to see which of the three variables seems to be causing the others, but we reached no firm conclusion.

We concluded that all three variables—education, intelligence, and internal control—were influenced by childhood factors. The most plausible conclusion would be that high intelligence causes a high sense of control, and that intelligence and internal control both influence the subject to obtain more education. But there is no evidence for this hypothesis.

People with high internal control are more likely to be northern-born and raised in a stable home where both parents were well educated. Why should these factors make a difference? In a way, this is like asking what it is about being middle-class that makes one middle-class, since a high level of internal control, with the attendant ability to defer gratification, to plan and save, and to control aggression, is the central hallmark of what we call being middle-class.

Why do middle-class parents (or more accurately in this case, stable working-class parents) have middle-class children? Many educators and social scientists have tried to understand "the hidden curriculum of the middle-class home."[13] We will not take space here to review all they have written. Our best guess is that well-educated parents are able to raise children in an environment that is not frightening—where discipline is less harsh, physical deprivation less common, and where the parent's success at living (and their higher level of internal control) presents a viable model for the children to follow. The children grow up in a household which teaches them, by both word and example, to defer gratification; where life is not so intolerable that one cannot put off gratification; where the authority figures are not harsh and unreasonable; where one learns to deal with frustration with explanation instead of fury. The higher internal control of children of better-educated parents is not simply a matter of increased education, although children of well-educated parents do have more education.

Men from broken homes do not have lower levels of education. One reason for this is that broken homes are more common among northern-born respondents whose educational opportunities are greater. Despite their southern backgrounds, however, boys from stable homes have higher internal control. We do not know for certain why this should be, but the fact that this association is consistently true only for men suggests that it may have to do with the absence of the male role-model. For women, there apparently is no relationship between family stability and internal control, but girls from stable homes are more likely to finish high school—a pattern exactly opposite that of men.

Northern-born respondents of both sexes have higher internal control. Only 12% of northern-born male high school graduates score low in internal control, compared to 33% of men who did not finish a southern high school.

---

[13]The quotation is from Fred L. Strodtbeck (1965), "The Hidden Curriculum of the Middle-Class Home" in John Krumbaltz, (ed.) *Learning and the Educational Process,* pp. 91-111, Rand McNally, Chicago.

Inhibition of Anger and Internal Control

So far we have developed two personality traits which seem to be associated with some elements of achievement: the ability to express anger without committing aggressive acts, and a sense of internal control. We have already seen that people with a sense of internal control are less likely to fight or be arrested; what about their ability to express aggression?

In analyzing inhibition of aggression, we saw that Southerners and the poorly educated were likely to be overly inhibited. We also saw that inhibition of aggression seemed to cause cautious, passive sorts of behavior and was associated with low achievement. From what we have seen so far, we would predict that those people who have a strong sense of internal control, who are from the North, well-educated, and high in achievement, would be uninhibited in their expression of anger. In fact, this is not true. Persons with high control are more inhibited, as Table 5.8 indicates.

TABLE 5.8
*Internal Control and response to "Do you
remember being angry?"*

Internal control	Unable to recall anger (%)	
	Men	Women
High	43 (616)	33 (640)
Medium	38 (868)	27 (1034)
Low	31 (364)	24 (544)
Gammas	.13	.13

The result is most unusual. Normally we expect two traits, both of which predict achievement, to go together. But in this case we have seen that the better-educated respondents are more likely to have a high sense of internal control, or else they are more likely to be able to express anger; but they are not likely to be able to do both. Passive optimism and aggressive fatalism are the competing alternatives.

What is a plausible explanation? We think that inhibition of aggression causes the respondent to express a high sense of internal control. People who inhibit aggression are generally uncritical; they have less antiwhite sentiment, for example. It seems that since they are unable to express anger, they avoid frustration by denying its existence. Saying that one's plans are subject to good and

TABLE 5.9
*Percent who can name an employer where they could
get a job, by Internal Control and recall of anger*

	Know of a job (%)	
Internal control	Recall anger	No recall of anger
MALES		
Low	34 (541)	16 (282)
High	41 (603)	25 (428)
FEMALES		
Low	22 (819)	16 (269)
High	36 (758)	22 (345)

bad luck is an example of recognizing the possibility of frustration.[14] Therefore
the inhibited person is likely to claim to the interviewer (and to himself) that he
lives in the best of all possible worlds. He may not, in fact, have a deep commitment to deferring gratification or organizing his future at all. At the very least,
his self-confidence is limited in that it cannot be expressed except in passive ways.
This is virtually impossible to demonstrate with survey data since the respondent
whose aggression is this completely inhibited will give very few answers suggesting that he has difficulty of any sort.

[14] Again one might raise the question of whether we are still correct in interpreting the
inability to recall being angry as inhibition of aggression. It might simply mean that the high-
control person is subject to less frustration, rather than denying it or internalizing aggression.
One scrap of evidence pointing in this direction is the relatively low education of the men
with high internal control who do not recall anger: only 49% are high school graduates,
compared to 69% of the high control men who do recall being angry. But more interestingly,
if we look only at non-high-school graduates we see that the high-control, non-angry group
are very likely to report having been in a fight. The following table demonstrates this rather
pretty interaction effect:

Percent in Fight (less than 12 years education, men only)

Control	Recall anger	Cannot recall anger
Low	59% (173)	26% (99)
Medium	47% (270)	32% (200)
High	39% (114)	48% (133)

The low-control, expressed anger group has a 58% fight rate, appropriate to this expressive
and rebellious category. But it is the inhibited high-control group which has the next highest
fight rate, 48%, reflecting the explosive level of internal aggression here. The low-control,
no-anger group, with a very low fight rate, is the group which seems most passive; while the
express-anger, high-control group is probably least frustrated.

The inhibited person with high internal control describes himself as happy and successful. But one item suggests that this is not a completely accurate self-estimate. As Table 5.9 indicates, he does not know where he could find another job now. Whereas 34% of the low-control men who can recall being angry say they know where they could go to work, only 25% of the inhibited high-control men do. If high control is purchased by inhibition of anger, this is one case where one would be better off without it.

What does this add up to? It means we must sharply qualify our finding that control of environment is an achievement-producing trait. It does indeed lead to a certain kind of achievement, but on the whole that achievement is the result of passive actions—*not* spending money, *not* quitting one's job, *not* getting into trouble. This passivity is reflected in the responses to the following question, which was asked of the parents in the sample:

> For every parent there are times when you enjoy your children more than other times. Here is a list of different things about children which parents have told us they enjoyed. I would like you to look at this list and tell me the two things which you have found nicest about little children.[15]
> 1. When they listen to what you tell them to do.
> 2. When they are clean and neat.
> 3. When they are polite and well-behaved.
> 4. When they hug and kiss you.
> 5. Playing with them.
> 6. When someone else tells you how smart the child is.
> 7. When they learn to do something new by themselves.

The respondents overwhelmingly select the two obedience items: 47% choose "when they listen to what you tell them to do," and 52% choose "when they are polite and well-behaved." This points out the excessive pressure black parents feel to keep their children "in line."

An additional 22% choose "when they are clean and neat"—another response which indicates satisfaction with rigidly controlled children. The two responses which deal with affection—"when they hug and kiss you" and "playing with them"—are each selected by 11%. Finally, 12% value praise from others about the child's achievements—"when someone tells you how smart the child is"—and 33% like "when they learn to do something by themselves."

When we contrast the answers given by respondents with different educational levels, we get an interesting picture of the different reward structures of the middle-class and lower-class homes. Men and women who are high school grad-

---

[15]This item is adapted from Gerald Gurin, Joseph Veroff, and Sheila Feld (1960), *Americans View Their Mental Health,* Basic Books, New York.

uates place much lower value on orderliness. They are slightly more likely to answer "when they are polite and well-behaved," but are much less likely to say "when they listen" or "when they are clean." They do not demand affection— they don't check "hug and kiss" as often as poorly educated respondents–but they do enjoy giving pleasure: they are twice as likely to check "playing with them." Similarly, they do not demand that their children bring credit to them, but they do enjoy their achievements and reward their independence: they check "learning to do something new by themselves" and do not give the other achievement answer, "when someone else tells you how smart the child is." In short, they make fewer demands on their children, are more permissive, value independence and achievement more; they want to give rather than take.

These findings come as no surprise. Twenty years ago, folk wisdom held that the lower classes were more permissive, but social science has shown repeatedly that the child-rearing habits of the poor are more rigid and demanding. In this set of questions, the poorly educated respondents appear selfish—they want their children to be obedient and loving and "be a credit to the family." It is only the middle-class parents who have the freedom to appreciate playing with their children or who can be more concerned with what the children learn than with the trouble they cause. Working-class parents do not say they like their children when they learn something new; they do say they like it when others say they are smart, which we think reflects the discontinuity between their aspirations for their children and their day-to-day child-rearing practices. They would like their children to achieve, but cannot encourage them in this direction; they seem more concerned with the childrens' impact on other people.

Well-educated parents usually have high internal control, so we should expect the pattern of associations to be similar, and in general they are. There are some interesting differences however.

Respondents with high internal control check the "learning by themselves" item, but not much more often than the low-control respondents, and they do not devalue "when someone else tells you how smart the child is" or "when they listen." Like the well-educated respondents, they value playing with their children and devalue being clean and neat. The suggestion is that high-control parents are affectionate and relaxed but do not stress independence as much. They stress achievement, but tempered by obedience and the ability to make a good impression.

There was certainly a time not long ago when this kind of passive and conformist achievement orientation was the only viable alternative for the upwardly mobile black. And even this kind of orientation is still difficult enough to sustain, for blacks do have lower internal control than whites. But it is an achievement orientation which, when coupled with passivity, leads only to the lower middle-class—toward being a postal clerk rather than an engineer.

Summary

Internal control is a measure of the degree to which an individual feels he has the power to control his environment. The data have shown that having "high" internal control is associated with inhibiting aggression, planning for the future, and being generally "middle-class" in one's attitudes toward achievement. In addition, people who possess this trait tend to come from stable, high-status backgrounds, suggesting that this strategy of adaptation to "the system" is learned while one is growing up. High internal control is also related to the trappings of success: high education, high income, job and marital stability. Hence internal control is a complicated concept, and having "high" internal control implies a rather ambiguous situation. On the one hand, it involves the inhibition of both anger and negative racial attitudes, but on the other, it provides a personality skill that is apparently necessary for success.

6
Self-Esteem

You learn more around white people. They are always discussing things. Learning new things. Negroes don't want to be bothered. They won't even watch TV unless it's something with another Negro in it.[1]

Self-esteem is an intriguing concept in an analysis of the social psychology of race. Minority group self-esteem has been widely discussed by both psychologists and sociologists, and the assumption is usually made that blacks "internalize" society's negative opinion of them and therefore cannot help but have lower self-esteem than whites. Sometimes data have borne this out, sometimes not, and this is due, at least in part, to the fact that numerous different measures of self-esteem have been employed and these have had varying reliability. Since there is no direct means of validating a measure of self-esteem, such a measure is usually evaluated on the basis of face validity, but there is no standard conceptualization of what self-esteem "is." In this chapter we will present a self-esteem scale, discuss the concept of self-esteem it is intended to measure, and study its relationship to anger and internal control for blacks and whites.

The interviewer approached the topic by saying:

Now I would like you to rate yourself as above average, about average, or below average on some things that you do and some things that you are.[2]

[1] A 35-year-old Detroit man, born in rural Pennsylvania.
[2] This item was drafted by Alice S. Rossi.

A list of ten items was then read, as shown in Table 6.1.[3] Notice that the items fall into three general groups. The first three items measure how the respondent feels about himself as a family member—in his roles as a child, a parent, or a spouse. The next two items ask him to rate himself as above average, average, or below average in two aspects of his character: trustworthiness and "willingness to work hard." The last items are ratings of his ability in five areas.

The table gives only the above average response. There is apparently great reluctance to rate oneself below average on any of the first six items; no more than 3% of the respondents rate themselves low as a child, parent, spouse, or in willingness to work, trustworthiness, or intelligence. However it is apparently acceptable to be below average in mechanical or athletic ability, as a conversationalist, or in one's sex appeal. For example, 41% of the white women rate themselves below average in mechanical ability, and 46% of the black women rate themselves low in athletic ability. White men rate themselves low in sex appeal (15%) and conversational ability (19%) more often than the other groups.

TABLE 6.1
Race and Sex Differences in Self-Esteem Items

Respondents rating themselves above average (%)	White male %[a]	Black male %[a]	White female %[a]	Black female %[a]	Mean differences in response:[b]	
					White-black	Male-female
As a son/daughter	21	20	19	19	0	1
Father/mother	24	22	26	28	0	−4
Husband/wife	28	26	23	27	−1	2
Trustworthiness	61	43	54	42	15	4
Willingness to work hard	59	44	49	40	12	7
Intelligence	34	18	18	16	9	9
Mechanical ability	40	29	18	14	7	18
Athletics	23	19	18	8	7	8
As a conversationalist	26	21	20	22	2	2
In your appeal to (women/men)	14	20	8	11	−4	7

[a]N's (Father/mother) = 208, 1269, 322, 1836; (Husband/wife) = 238, 1238, 314, 1372; and (All others) = 283, 1855, 367, 2231. White N's are unweighted.
[b]Positive numbers indicate that whites or men are more likely to rate themselves above average.

[3] See Appendix 2 for analysis of the scalability of the self-esteem items.

On most of the items, whites are more likely than blacks to say they are above average. The fifth column of Table 6.1 gives the average racial difference. For example, in "willingness to work" white men are 15% more likely to rate themselves above average (59 vs. 44%) and white women, 9% more likely. This averages out to a 12% difference. The sixth column shows that men rate themselves higher than do women. But there are differences among the items. The three dealing with family relations show no race or sex differences at all—approximately one-fourth of each group see themselves as above average. On the character items, over half of the whites rate themselves as above average in trustworthiness and willingness to work hard—two highly valued character traits. White women are slightly less favorable to themselves, and black men and women are less likely than whites of their sex to rate themselves above average.

Turning now to the ability items, intelligence shows an interesting pattern. A third of the white males rate themselves above average, but white women and blacks both rated themselves much lower. Black women rated themselves lowest of all. White males are twice as likely as any other group to consider themselves smarter than other people.

Mechanical ability shows a sharp racial difference, with black men considerably less likely to rate themselves above average. Women rated themselves below average on this item as often as they did on athletics.

The stereotyped black is glib and sexy, and apparently blacks believe the stereotype. Black women rate themselves higher than white women in conversational ability, and blacks of both sexes rate themselves above whites in sex appeal. A significant number of people, especially whites, rated themselves below average on these two items. In view of these results, it is surprising that blacks do not rate themselves higher in athletics, but the table shows that whites are more likely to call themselves above average. In fact, black males rate themselves nearly the same as white females rate themselves on this item.

These differences between men and women lend support to Gurin's observation that women and men organize their self-perceptions around those factors that are relevant to their social roles:

> Women more than men will turn within themselves for identity anchors. A man looks for sources of self-esteem in external accomplishments, whereas a woman finds her sources of self-esteem in her internal responses and feelings . . . . We would also expect her to stress the interpersonal in looking at her relation to the world.[4]

This is not the whole story, however, for this does not explain why women rate themselves lower on items like trustworthiness. Apparently women *do* have

[4]Gerald Gurin, Joseph Veroff, and Sheila Feld (1960), *Americans View Their Mental Health,* p. 78, Basic Books, New York.

generally lower self-esteem than men, and the differentiating items are ability and character items. The black self-esteem ratings are appreciably lower than those of whites on four items—mechanical and intellectual ability, and the critical virtues of trustworthiness and willingness to work.

The average white male rated himself above average on 3.3 items; white females, on 2.5; black males, on 2.6; and black females, on 2.2. Relative to men, black women do not rate themselves as low as do white women; black women have higher esteem relative to black men than white women do relative to white men.

According to this self-esteem scale, blacks have, on the average, lower self-esteem than whites, but the racial differences are not so straightforward. There is more "spread" in the distribution of self-esteem scores for blacks, and blacks tend to have either very low or very high self-esteem. While 27% of the black respondents refused to rate themselves above average on any item, only 15% of the whites rated themselves this poorly; in addition, more blacks than whites give themselves an "above average" rating on seven or more items. The reason for this pattern of black self-esteem scores has to do with region of birth. While 50% of all whites score higher than 2.0 on the self-esteem scale, only 41% of all blacks score that high; but while 35% of southern-born blacks score higher than 2.0, 59% of northern-born blacks score that high. Thus blacks born in the South have generally lower self-esteem than whites, but blacks born in the North tend to have self-esteem as high or higher than whites. Hence the standard assumption that blacks necessarily have lower self-esteem than whites must be qualified when this particular measure of self-esteem is used.[5]

The Self-Esteem Items as a Scale

Although the ten items refer to widely different traits, they do fit together to measure self-esteem. The person who rates himself above average in one area is likely to rate himself above average in others. The gammas between all possible pairs of the ten items are consistently positive.[6] The gammas are relatively weak in associations with mechanical ability and athletic ability. This result is due to the fact that these items do not work for women, regardless of race. Women are not expected to be athletic or mechanical, and these characteristics are therefore of small importance to female self-esteem. The remaining γ's are consistently high, usually above .40.

---

[5] Similar findings have appeared in studies using other measures of self-esteem. See, for example, E. Earl Baughman and W. Grant Dahlstrom (1968), *Negro and White Children,* Academic, New York; and E. Earl Baughman (1971), *Black Americans,* Academic, New York.

[6] The matrix of gammas and a factor analysis of the scale are given in Appendix 2.

There is only a slight tendency for the items to cluster into subscales. For example, the gammas between evaluation of oneself as a son or daughter and as a father or mother are .54 for blacks and .58 for whites. These are high, but no higher than the association between son or daughter and intelligence for blacks (γ = .56) or son or daughter and trustworthiness for whites (γ = .61). Apparently the questionnaire is a measure of a general sense of self-pride rather than a purely objective evaluation.

Indeed it seems unlikely that the respondents are merely reporting the objective truth about their ability. It is hard to see how one could know whether he or she was more trustworthy, or a better mother, than average. One area where one might have a basis for an objective judgment is in intelligence, and the respondents' self-ratings of intelligence correlate only moderately well with the verbal test score (γ = .21 for men and .29 for women). If one were trying to predict whether someone would rate himself above average in intelligence, one would have a better chance knowing his self-rating as a parent than knowing his actual I.Q.

The Meaning of Self-Esteem

It seems that self-esteem does measure a psychological trait, since the correlations between the items are high, and the scale does not seem to be a mere reflection of objective reality. But what *is* "self-esteem?"

Attempts to define and measure the notion of "self-concept" or "self-image" are numerous in the literature. In the tradition of Cooley's "looking-glass self" and Mead's "Me," sociologists tend to emphasize the social aspect of the self, or the self as a reflection of society's norms and values.[7] Cooley's conception of the self-image emphasized the individual's *perception* of how others think of him. Mead's "Me" represents that aspect of the self which incorporates the attitudes and expectations of the social group (the "generalized other"). Implicit in these approaches is the assumption that self-images cannot exist except within the social process; "others" must exist in order for self-perceptions to exist.

The concept of self-esteem that is measured in the present study draws on the experience of several researchers in this area and emphasizes self-evaluations. Gurin *et al.*[8] measured perceptions of the self in terms of the person's willingness to define himself as a unique individual with traits different from those of everybody else. Three questions asked respondents to express "ways in which

---

[7] These are the classical sociological works on the "self," but generally lack empirical verification: Charles Horton Cooley (1930), *Human Nature and the Social Order*, Scribner, New York; George Herbert Mead (1934), in *Mind, Self, and Society*, Charles W. Morris (ed.), Univ. of Chicago Press, Chicago.

[8] Gurin *et al.*, *op. cit.*

you are different from other people," how they would like their children to be different from themselves, and what their "good points" and "strongest points" are.[9] Logically, uniqueness can refer to perceptions of being positively unique or negatively unique. Rosenberg focused instead on emphasizing *satisfaction* with the self.[10] In Rosenberg's measure, "high" self-esteem is indicated by strong feelings of self-satisfaction, and this implies no sense of having a unique identity:

> High self-esteem, as reflected in our scale items, expresses the feeling that one is "good enough." The individual simply feels that he is a person of worth; he respects himself for what he is, but he does not stand in awe of himself nor does he expect others to stand in awe of him. He does *not* necessarily consider himself superior to others.[11]

The self-ratings used in our scale seem to be a direct way to measure self-esteem, and the scale simplifies matters by defining for the respondent the ten areas of interest. Yet our conception of self-esteem is different from either Gurin's or Rosenberg's conceptions, since it involves more directly the notion of self-evaluation. Our questionnaire, building on Gurin, asks the respondent to see himself as not only different from the average person, but in some ways superior. To say that one is above average in various ways is very different from merely expressing self-satisfaction or self-acceptance. For Rosenberg, a respondent with high self-esteem need only say "there is nothing wrong with me" or "I like me." For Gurin, the respondent must say "I am a unique individual." But for our measure, he must say "I am unique—a good and talented person, better than the average person in some ways."

All of these complexities may explain why self-esteem is an elusive concept. It is probably the traditionl "social" orientation to self-esteem which has persuaded almost all sociologists and psychologists to assume that blacks have low self-esteem; they could not help but accept the view of themselves held by their "significant others"—whites.[12] But self-esteem, as we have measured it, has a tone of competing against others, rather than merely accepting the views of others toward oneself.

[9]*Ibid.,* p. 53.

[10]Morris Rosenberg (1965), *Society and the Adolescent Self-Image,* Princeton Univ. Press, Princeton, N. J.

[11]*Ibid.,* 31.

[12]The notion that whites are "significant others" for blacks has been called into question by several researchers, as noted by J. D. McCarthy and W. L. Yancey (Jan. 1971), "Uncle Tom and Mr. Charlie: Metaphysical Pathos in the Study of Racism and Personal Disorganization," *Am. J. Soc.,* **76** (No. 4), 648-672.

The Correlates of Self-Esteem

How can we interpret the willingness to perceive oneself as "better than" other people? High self-esteem and recalling anger are associated; this suggests that having high self-esteem is an indicator of a trait we will call "assertiveness." Assertiveness refers to competitiveness and the tendency to manipulate the environment to one's own advantage. Assertive people are socially aggressive: they seek self-improvement and actively take steps to assure their well-being. The data include several variables that indicate assertiveness: having a checking account, knowing where one can find a job, participating in civil rights demonstrations, being concerned about integration. Table 6.2 shows that self-esteem is positively correlated with all these indicators of assertiveness for blacks.

Of course, persons with high esteem are better educated, as Table 6.2 shows, but the right-hand half of the table shows that education is *not* responsible for the associations, which are not eliminated when education is controlled. This suggests that self-esteem is a personality trait that is, at least in some ways, conducive to achievement. In fact, education may itself be caused by self-esteem;

TABLE 6.2
Association between Self-Esteem and Assertiveness by Sex (γ)

	Males	Females	Males		Females	
			Low education	High education	Low education	High education
1. Thinking about getting a new job	+ .22	+ .04[a]	+ .18	+ .16	− .07[a]	− .02[a]
2. Know where to get job	+ .30	−	+ .07	+ .37	−	−
3. Took vocational training in high school	+ .18	+ .16	+ .11	+ .11	+ .14	+ .17
4. Would pioneer in white neighborhood	+ .15	+ .15	+ .05	+ .20	+ .09	+ .25
5. Prefer integrated neighborhood	+ .30	+ .09	+ .17	+ .15	− .19	+ .11
6. Present contact with whites	+ .20	+ .29	+ .17	+ .18	+ .22	+ .30
7. Took part in civil rights demonstrations	+ .19	+ .24	+ .06	+ .22	− .07	+ .30
8. Have checking account	+ .17	+ .27	+ .13	− .02	+ .25	+ .15
9. Education	+ .32	+ .18	−	−	−	−
10. Income	+ .08	−	+ .09	+ .03	−	−

[a]Gamma refers only to those women who work.

seeking higher education may be interpreted as another example of the assertiveness of high-esteem respondents. Having high self-esteem is also associated with the desire to be integrated (i.e., preferring a white neighborhood, willing to be the first to move into a white neighborhood, participating in civil rights demonstrations), and with the manipulation of resources (i.e., job betterment, financial security, educational preparations). Furthermore, self-esteem is positively associated with parents' education, especially the mother's education, indicating that parents may transmit a sense of self-assertiveness to their children. Another way of saying this is that people *learn* to be assertive.[13]

This finding is similar to one of Rosenberg's: that high self-esteem (as he measured it) corresponds to a preference for competitive work situations and to the expectation of getting ahead in life.[14] According to Rosenberg, people with low self-esteem expect frustration and defeat in competitive situations, whereas people with high self-esteem expect to have "above average" success. Assertiveness is therefore associated with an individual's expectations of personal success.

High self-esteem is also associated with the dysfunctional aspects of expressing aggression. Our data show that people with high self-esteem openly express anger more than people with low esteem, who presumably internalize their anger (i.e., turn it against themselves). Table 6.3 shows that blacks with high self-esteem tend to get into fights, get arrested, and be separated or divorced more often than people with low self-esteem. These associations hold, at least for men, when education is controlled, so that the effect of education (or social class) is not responsible for the relationship between self-esteem and behavior. These findings demonstrate that Kardiner and Ovesey's model of black personality was overly simplistic in assuming that *low* self-esteem is associated with aggressive behavior, behavior which serves to compensate for unfavorable evaluations of the self.[15]

The right-hand part of Table 6.3 shows that for white men and women who did not finish high school, those who have fought or been arrested have low self-esteem. This is of course the opposite of the black pattern. The person who commits deviant acts and has low self-esteem seems in some important way different from the deviant with high self-esteem. Our impression is that the deviant with low self-esteem is the defeated person, the drifter, who gets into trouble partly because he has no real reason to stay out of trouble. On the other hand, the image which comes to mind of the deviant with high self-esteem is the image of the aggressive person, pushing to dominate others and getting into trouble because he is "too ambitious." The few white high school graduates who have fought or been arrested have high self-esteem.

---

[13]See Chapter 10 for a discussion of the causes of self-esteem.
[14]Rosenberg, *op. cit.*, p. 227.
[15]Abram Kardiner and Lionel Ovesey (1951), *The Mark of Oppression*, Norton, New York.

TABLE 6.3
Association between Self-Esteem and Expression of Anger

| | Blacks (γ) | | | | Whites (γ) | | | |
| | Male | | Female | | Male | | Female | |
	Low education	High education	Low education	High education	Low education	High education	Low education	High education
Been in fights	+.23	+.27	−.06	−.05	−.49	+.24	−.24	+.17
Been arrested	+.12	+.07	+.03	+.43	−.41	+.25	−.35	−.04
Recall last time angry	+.14	+.26	+.05	+.13	—	—	—	—
Separated or divorced	+.15	+.14	−.02	+.11	—	—	—	—

Now we may be in a position to understand why region of birth is so strongly associated with self-esteem, and why violence may be a more serious problem in northern ghettos than it is in the South. Self-esteem for men is more strongly related to region of birth than it is to either education or income. (For men, γ = .46, compared to .32 for education, and .08 for income). This is one of the largest gammas in the analysis. We think this is related to the fact that Northerners can express anger much more easily than Southerners.

Perhaps southern blacks have lower self-esteem than Northerners because the South is more blatantly discriminatory than the North. Blacks in the South are socialized to be passive and obedient in order to survive. In the North, on the other hand, modes of proper behavior for blacks are less well-defined since discrimination is likely to be in more subtle forms. Blacks in the North can afford to express frustrations more openly. And it seems likely that being able to express anger and frustration makes it easier to say, in effect, "I *am* better than other people. It's not my fault that I can't succeed!"

Self-Esteem and Internal Control

So far we have shown that our measure of self-esteem taps an element of behavior we have called assertiveness. It is associated with education and such "middle-class" attitudes and behavior as knowing where to find jobs, wanting integration, having a checking account. The internal control scale is also a measure of middle-class behavior; persons with high internal control have more education and income, save their money, own their homes, etc.

We might expect that these two variables measure much the same thing and should therefore distinguish two main character types: those with low self-esteem and low internal control and those with high self-esteem and high internal control. We would expect the former group to be essentially lower-class people and the latter group to be middle class. Instead, we find that the association between internal control and self-esteem is very weak for blacks when the effect of education is removed, but remains strong for whites (see Table 6.4). This means that for blacks there are four distinct character types, as defined in terms of high and low self-esteem and high and low internal control, and that all four types are equally likely (i.e., equal proportions of the sample fall into each of the four types). We can still be certain, however, that the social class factor exerts pressure for internal control and self-esteem to be consistant, since, for example, increasing education raises both of these. It follows, therefore, that there must be some other factor operating which forces internal control and self-esteem to diverge for blacks. To some extent, for blacks, internal control and self-esteem must be contradictory characteristics, so that the presence of one tends to exclude the other to a degree.

TABLE 6.4
Internal Control and Self-Esteem by Sex and Education

	Association between internal control and self-esteem (γ)	
Education	Blacks	Whites
MALES		
Low	+ .10	+ .27
High	+ .03	+ .16
FEMALES		
Low	− .01	+ .25
High	+ .19	+ .35

The different combinations of self-esteem and internal control represent alternative strategies in two ways: (1) they depict a choice between externalizing anger (system-blaming) and internalizing anger (self-blaming); and (2) they represent a choice between assertive behavior and deferential or passive behavior. Thus, for example, a person with high self-esteem and high internal control would be expected to be assertive (competitive, manipulative) and to believe that he controls his destiny. We will call this type of person an "achiever." A person with low internal control and low self-esteem (a "drifter") will be noncompetitive but angry or aggressive. The two other types, "accepters" and "militants," represent more complicated adaptations. Accepters know how to achieve but are passive. Militants are competitive and assertive, but without any sense of being able to obtain their goals. The four types represent a typology in which the major differences between alternative types lie in the handling of aggression and they depict alternative adaptations to discrimination for blacks.

	Internal Control	
	Low	High
Low self-esteem	Drifters	Accepters
High self-esteem	Militants	Achievers

The conceptual scheme represented by the typology has been implied by at least one other researcher. Caplan[16] has studied the results of several studies of

[16]Nathan Caplan (1970), "Identity in Transition: An Empirically Based Theory of Black Militancy" in *Revolution Reconsidered*, N. Miller and R. Aya (eds.), The Free Press, New York. It is interesting that Caplan hypothesizes internal control as part of his "personal efficacy" dimension; however, it appears that the two studies he cites which use an internal control scale also define militancy in a different manner, so that this seeming disagreement can be rejected.

riot participation and militancy among blacks and has found some general trends in the data. He claims that "militancy" occurs in those individuals who combine a strong sense of self-worth (which he refers to as "personal efficacy") and a low sense of environmental control (which he calls "political efficacy"). Caplan's "militant" category, which includes those identified as rioters, are clearly related to our militant type, both psychologically and behaviorally. Caplan cannot present statistical identification of four types, but his recognition of two independent dimensions of efficacy provides some evidence of the soundness of our typology.

Thus we have four strategies for achievement. Which is the best one? The achievers are the best educated, and the drifters have the least education, as Table 6.5 indicates; the accepter and the militant fall in between. For women and older men, accepters are better educated than are militants; but for younger men, militants are the second-best educated group.

The achievers, who are both older and better educated, should have much higher incomes, and they do. However, even when differences in age and education are controlled, male achievers still have incomes over $300 per year higher than the militants. The militants in turn have incomes about $200 per year higher than either accepters or drifters. Obviously those who are high in both internal control and self-esteem are most successful, and those low in both characteristics, the least. But if one must choose between high esteem and low control, or high control and low esteem, the choice is difficult; the militant has the higher income, the accepter the better education. Apparently both strategies have strengths, and both weaknesses.

The weakness of the militant is, of course, the handling of aggression. This is shown in Table 6.6. The percentages in the table are averages computed across four groups within each type: low education males, high education males, low education females, and high education females. For example, drifters include people from each of these four groups, although not in equal proportions. By presenting the simple

TABLE 6.5
Self-Esteem, Internal Control, and Education (controlling on sex and age)

	Graduated from high school (%)			
	Men		Women	
	21–29	34–45	21–29	30–45
High control, high esteem (achievers)	73 (219)	60 (274)	71 (149)	69 (290)
High control, low esteem (accepters)	62 (173)	44 (348)	60 (273)	51 (405)
Low control, high esteem (militants)	73 (139)	22 (211)	47 (175)	30 (224)
Low control, low esteem (drifters)	41 (150)	19 (306)	42 (259)	25 (446)

TABLE 6.6
Consequences of the Typology^a

	Percent having trait named in row			
	Low S E low I C (Drifters) (%)	Low S E high I C (Accepters) (%)	High S E low I C (Militants) (%)	High S E high I C (Achievers) (%)
Aggression				
Ever in fight	30	21*	30	*35*
Ever arrested	27	17*	*28*	23
Recall last time angry	70	59*	*75*	65
Ever divorced, separated	33	30*	*38*	32
High anti-white	32	19*	*33*	21
High anti-black	69	60*	*70*	64
Ever evicted	12	7*	*14*	9
Discontent				
Do not trust people	*74*	63*	69	63*
Unhappy	29	14	*34*	13*
Dissatisfied with job	*28*	14*	23	16
Civil rights				
Organizational member	14*	21	19	*29*
Has demonstrated	27*	31	36	*37*
N^b	(1018)	(605)	(987)	(745)

^{a}Asterisk indicates smallest percentage in row; highest number in italic type. The data are averages for four education-sex categories.
^{b}N's refer to the smallest N for that type in the given set of variables.

average of the four education-sex combinations, it is possible to see what percentage of drifters would have a certain trait if the drifters had equal portions of men and women with high and low education. This means that the differences between drifters, accepters, militants, and achievers are the differences remaining after the education and sex differences have been removed.[17]

The dependent variables in Table 6.6 are divided into three groups: those that refer to handling aggression, those that indicate general discontent, and those that demonstrate an interest in civil rights activities. Looking at the aggression group first, we can see that drifters and militants exhibit the higher rates of aggressive behavior, and accepters exhibit the lowest. For example, 27% of the drifters and 28% of the militants have been arrested, but only 17% of the accepters and 23% of the achievers have been arrested. These figures represent a percentage difference of 10 points between the rate of accepters and those of drift-

[17]This procedure is called "direct standardization."

ers and militants. In addition, militants and drifters are more likely than accepters and achievers to express hostile attitudes. The table shows that militants and drifters are more anti-white and more anti-black than either accepters or achievers. All of the aggression variables support this trend; the accepters are the most passive group, and the militants and drifters the most aggressive.

The discontent variables generally repeat the same pattern as the aggression variables, but in addition they differentiate the achievers from the drifters and militants. The achievers are as contented as the accepters. The drifters and militants are the most discontented; they are the most unhappy and frustrated, are dissatisfied with their jobs, and are more likely not to trust people. The accepters and achievers, on the other hand, exhibit the same low rates of discontent.

Regarding civil rights activities, a new pattern emerges. Achievers are the most active, drifters the least. Militants are as likely as achievers to participate in demonstrations, but not in organizational activity. This seems to be a further reflection of the militants' aggressiveness.

To sum up Table 6.6., we can say that the typology differentiates people along these three dimensions at least. In terms of aggressiveness, militants and drifters are more expressive of anger and hostility than the other types. Accepters are the most passive type, and achievers are as unlikely as accepters to express discontent to the interviewer. For the achievers, there seems to be some inconsistency between attitudes and behavior, since their lack of discontent does not seem to keep them out of fights, for example. But their high rate of participation in civil rights activities may suggest an explanation; they are not complainers, they are doers.

In the preceding chapter, we saw that for many respondents high internal control was the result of excessive inhibition of aggression. Internal control was a qualified virtue. In this chapter we see a similar pattern with self-esteem. The high self-esteem subject who has a weak sense of internal control focuses his anger at "the system" (or more correctly the environment, including other blacks). He thinks highly of himself, but blames the world for failing him—he is the most frustrated and unhappy of the four groups. Thus, being able to live successfully seems to require having both high self-esteem and high control. Unfortunately this combination is uncommon; high esteem and high control do not naturally accompany each other.

Perhaps the most interesting fact to emerge from this analysis is that the typology is senseless when applied to whites. Whites in our sample tend to be one of two types: low self-esteem and low internal control or high self-esteem and high internal control. The behavior variables for whites are related to social class factors and not to the psychological variables comprising the typology. What makes blacks different?

In Chapter 10, we will look at the causes of both self-esteem and internal control, and will get a better understanding of the forces that cause control and esteem to be incompatible. But it already seems clear that, as many other writers have said, aggression-handling is a critical factor. Faced with the fact of race prejudice, the black is under strong pressure either to reply with anger and become militant, blaming everyone but himself for his troubles, or to accept society's definition of the situation: that the world is fair and will give him a break as long as he remains passive and feels properly undeserving.

Summary

The self-esteem scale measures the degree to which an individual is willing to assert that he is better than other people. While it is true that, in general, blacks score lower on self-esteem than whites, northern-born blacks score as high as whites. This is not due to integration (see Chapter 10), however, but to the increased opportunity of expressing aggression in the North. Self-esteem is associated with a characteristic called "assertiveness," which is defined as functional aggressiveness (i.e., competitiveness that enables the individual to manipulate his environment successfully). Self-esteem is also associated with dysfunctional aggressiveness (i.e., fighting and getting arrested). The critical factor is the individual's sense of internal control, which seems to be independent of his self-esteem, at least for blacks. The data reveal four types of blacks who handle aggressiveness differently, depending on their self-esteem and sense of control.

7
Happiness Is...

We have begun to describe the dilemma facing blacks in the North. Scoring high on the internal control scale implies a healthy confidence in one's ability to master the environment, but if this strong sense of control is developed by inhibiting all hostile feelings about that environment, then having high control may mean denying reality. Similarly, people with high self-esteem are self-confident and assertive, but the data show that high self-esteem is often associated with overly aggressive, antisocial behavior. So the objective observer cannot say that one strategy or the other is "best" in its psychological or social benefits to the individual. If the observer cannot say which individuals are better off, what do the respondents say? How many blacks (and which ones) say they are happy?

We have used a question from Gurin, Veroff, and Feld's national survey of mental health[1] to measure happiness:

> Taking all things together, how would you say things are these days—would you say you are very happy, pretty happy, or not too happy?

Norman Bradburn[2] used the same question in two studies. The second Bradburn study was conducted by combining several different surveys, from different types of communities and different regions of the country. One of his communities was a ghetto in Detroit.

[1] Gerald Gurin, Joseph Veroff, and Sheila Feld (1960), *Americans View Their Mental Health,* Basic Books, New York.
[2] Norman M. Bradburn and David Caplovitz (1967, 1969), *Reports on Happiness,* NORC Monographs in Social Research, Aldine, Chicago; and *The Structure of Psychological Well-Being,* NORC Monographs in Social Research, Aldine, Chicago.

Table 7.1 shows the response to the happiness question in Gurin's survey, for the whites and blacks in Bradburn's second study, and for our respondents. The table shows consistent results, with proportions in the middle category nearly identical. The unhappiest are the blacks in the group interviewed by Bradburn. Our respondents are slightly more happy. Gurin's national sample, which is 90% white, is much happier, and Bradburn's group of whites slightly happier still. The racial differences are quite sharp; blacks are at least twice as likely to say they are "not too happy."

Of course, we don't know whether we have "really" measured happiness or not. Obviously a psychiatrist would want to probe much deeper before judging whether a respondent was "well-adjusted," and even then he might not be able to say whether he was happy. Perhaps, in the final analysis, our straightforward question is the best way to measure happiness; each individual must decide for himself whether he is happy. Nevertheless one should not read these data as meaning that most people are happy. As with all survey questions, the happiness question is a verbal stimulus, which elicits a verbal response that indicates simply the person's willingness to say he is happy. The very expression of such feelings is behavior, and like all behavior it is guided by the norms of the society. For example, people are expected to be happy and will be reluctant to admit they are not. Further, excess emotionalism is frowned upon by society. Both of these together mean that very few people will say they are unhappy. Even when a gentle euphemism like "not too happy" is used, only a small fraction of whites will choose this response, as Table 7.1 reveals. In addition, we should note that when a respondent is given three choices, he will generally tend to pick the middle one. Over half the respondents picked the middle category in every sample in the table. It might have been better had the question posed four response categories.

All this means we will have to be content with small differences in the tables that follow, and we must be cautious in interpreting them.

TABLE 7.1
Happiness in Three Surveys

Survey	Very happy (%)	Pretty happy (%)	Not too happy (%)	Total (%)	N
Gurin, et al.[a]	35	54	11	100	2460
Bradburn (whites)[b]	35	56	9	100	2219
Bradburn (blacks)[b]	18	57	25	100	516
Present study	23	55	22	100	4131

[a]Data from Gurin, Veroff, and Feld, *op. cit.*
[b]Data from Bradburn, *The Structure of Psychological Well-Being,* p. 45.

The Dimensions of Happiness: Gurin and Bradburn

In the Gurin survey, the happiness question was followed by others designed to probe for the sources of happiness and the sources of unhappiness. The most common source of both happiness and unhappiness was financial security. Other common sources of happiness were marriage, children, interpersonal relationships, and the respondent's job. Jobs and personal problems were given as causes of unhappiness as well. National issues were sometimes a cause of unhappiness, but never a cause of happiness.

In Gurin's analysis, we begin to see two factors emerging: first, the fact that happiness and unhappiness are not merely opposite sides of the same coin; secondly, that some of the causes of happiness and unhappiness are temporary "good things" which happen to the respondent, while others reflect a stable style of reacting to other people and the environment.

Bradburn argues that happiness is related to the balance between two statistically independent dimensions: positive and negative feelings. The person who has strong positive feelings does not necessarily have fewer negative feelings. If positive feelings dominate, the person is happy, according to his own self-report; if negative feelings dominate, he is unhappy. Positive feelings included feelings of pleasure in accomplishment, pride, interest in something, and feeling "on top of the world." Negative feelings included boredom, restlessness, loneliness, depression, and being "upset" by criticism. Bradburn also argues that self-reports of happiness are highly stable over time (that is, they have high test-retest reliability) indicating that happiness is in large part an enduring attitude toward life that is relatively unaffected by specific events. This last point is of special interest to us in the present study, since we argue that the happiness question measures a predisposition toward expressing overall psychological well-being or the absence of anxiety.

In addition, the Bradburn studies reveal a few basic relationships which the present study supports. In general, respondents with more education are happier, as are those with higher income, the married (as opposed to the widowed, divorced, or separated), the young, and the employed (for males).

If happiness does have both a stable and a fluctuating component, and if we should think of happiness and unhappiness as qualitatively different, then it begins to look as if the happiness response may break down into four components: (1) reactions to the favorable occurrences of everyday life; (2) the person's willingness to respond to positive occurrences; (3) reactions to the misfortunes which occur from time to time; and (4) the respondent's level of "internal" unhappiness—his degree of anxiety or depression.

In the tables that follow, we have grouped responses to form a dichotomy; the "happy" and "pretty happy" people as opposed to the "not too happy." Partly this is for simplicity, but also it is because the "not too happy" response is the one which runs clearly counter to social expectations, and the most definite (and hence most reliable) response. This procedure follows Bradburn's method of analysis.

Happiness and the Self-Esteem-Internal Control Typology

Let us begin by relating happiness to the self-esteem-internal control typology. Which of these two strategies for adjustment produces people who say they are happy?

Table 7.2 shows that accepters and achievers are the most happy, with 15 percentage points separating them from the drifters and 20 points separating them from militants. Clearly, happiness seems to be related to having high internal control but not to high self-esteem. Since better-educated respondents are noticeably happier ($\gamma = +.20$ for males and $+.17$ for females) we must control on educational attainment. When we do, we find that education is *not* responsible for the high association between internal control and happiness, however. For men, the gammas between internal control and happiness are $+.69$ for men who did not finish high school and $+.55$ for men who did; for women the gammas are $+.22$ and $+.36$. The gammas between self-esteem and happiness remain close to zero.

Happiness, then, seems to be measuring somewhat the same things as the internal control scale and could coincide with the inhibition factor of internal control. Since happiness is not related to self-esteem, the individual who says he is happy would appear to be one who is free from anxiety about "the system," regardless of his evaluations of himself. The happy person believes that he controls his destiny, and is not anxious about being treated unfairly by "luck" or "chance." This is especially interesting because it indicates that blacks with high internal control, who inhibit their feelings of frustration about the system, nevertheless say they are happier than people who do *not* inhibit their feelings. At least in terms of our measure of happiness, inhibition does not appear to be a psychologically depressing phenomenon.

Other indicators of inhibition bear this out. Table 7.3 shows the associations between being happy and being inhibited in terms of several other variables, some of which indicate a specific behavior (e.g. recalling a childhood incident of

TABLE 7.2
Happiness by the Internal Control Self-Esteem Typology
(% "not too happy")

	Low internal control	High internal control
Low self-esteem	Drifters: 29 (1162)	Accepters: 14 (749)
High self-esteem	Militants: 34 (1204)	Achievers: 13 (935)

TABLE 7.3
Happiness as a Measure of Inhibition

	Male (γ)[a]	Female (γ)[a]
Recall racial incident in childhood	−.23	−.12
Recall last time angry	−.26	−.23
Agrees to "sometimes I get so frustrated I could break things"	−.22	−.38
Anxious around white people	−.26	−.17
Want to "get even" with whites	−.31	−.15
Anti-white attitudes scale	−.29	−.18

[a]Gammas show associations between being happy and exhibiting the trait or behavior indicated. The negative direction of the gammas means that "happy" people do *not* recall racial incidents, say they get frustrated, etc.

racial discrimination) and some of which indicate the willingness to express an unpleasant attitude (e.g., anti-white feelings or a sense of frustration). Note that all gammas are negative. People who say they are happy are *less* likely than unhappy people to recall being racially discriminated against as a child, recall the last time they were angry, admit feeling frustrated or anxious around white people, or say they want to "get even" with white people. It appears, therefore, that the same people who are inhibited in expressing feelings of anger or frustration are also inhibited in expressing feelings of unhappiness.

One of the reasons why this is interesting is that a great deal of attention has been paid to the idea that inhibition of anger is a cause of depression. Supposedly depression is a common middle-class neurosis, and the association between inhibition and depression should show up in our data. But, for blacks, inhibition is associated with expressing happiness.

Happiness and Security

People who say they are happy are in some cases simply too inhibited to admit that things are going badly for them. But another reason why they seem inhibited is that they have less to complain about. Table 7.4 shows that "happy" people do have greater economic security, a more secure family relationship, are less afraid, and are more optimistic about civil rights. The table is consistent; all 14 gammas are positive, and 11 of the 14 are above .20. This agrees with Gurin and Bradburn, both of whom find strong associations between financial security and happiness.

Financial security is not simply income. In fact, income is the weakest predictor of happiness of all the variables in Table 7.4. And even this association between income and happiness does not support a simple "money buys happiness" theory. Table 7.5 plots happiness by family income, and shows that men and

TABLE 7.4
Happiness as a Measure of Freedom from Anxiety

	Male (γ)	Female (γ)
High family income (4500 or more)	+.20	+.13
Married (i.e., not widowed, separated, or divorced)	+.33	+.40
Satisfied with job	+.53	+.31
Can pay off debts	+.24	+.35
Never been evicted	+.46	+.31
Not afraid of being robbed	+.28	+.13
Believe that whites favor blacks' civil rights	+.27	+.03

women with family incomes above $4500 are considerably happier than those with lower incomes. Apparently $4500 (in 1966) was the breaking point between insecurity and security. But either above or below this point, income makes no difference. Persons with very low incomes are not more unhappy than those with incomes near $4000; persons with very high incomes are not happier than those making around $5000. The same pattern can be demonstrated using individual income instead of family income. Individual income is associated with happiness with a γ of .18 for men, .17 for working women. But suppose we again divide income at $4500 and ask, what is the relationship between happiness and individual income for incomes below $4500? The answer is that there is no association: $\gamma = .01$ for women, and $-.06$ for men—a slight tendency for the very poor to be happier than those with incomes near $4500. Similarly, above the $4500 line, income is again not associated with happiness for men ($\gamma = .01$), although it is for women—the association between income and happiness for women earning over $4500 is a respectable .18. Bradburn observed a similar (although not as pronounced) pattern for whites:

> It appears that severe income deprivation does have a strong relationship to happiness. Those who have considerably below-average incomes are likely to have a low

TABLE 7.5
Happiness by Family Income

Income ($)	% Happy		N	
	Male	Female	Male	Female
0–1500	69.6	79.2	89	198
1501–3500	69.7	74.0	205	351
3501–4500	65.4	76.8	240	164
4501–5500	84.3	83.8	332	99
5501–8500	80.0	89.0	698	119
8500 or more	84.3	84.6	185	26

sense of well-being. Beyond a certain income level, which empirically appears to be about $5,000 a year, the effect of further increment in income is moderate, although at most levels it continues to appear.[3]

Happiness As a Personality Trait

Thus far we have seen that security—the financial security of an adequate income, the absence of loneliness and fear—are important correlates of happiness for blacks. Happiness is associated with the absence of anger, and the presence of a sense of internal control. For men, these last two associations are extremely strong. All of this together seems to add up to a conception of the person who says he is happy as someone who feels that the world is a reasonably safe place— that there is no reason to be angry with one's fate.

But whether one is afraid or angry is partly a matter of personal choice. Some people, given a certain amount of security will be happy, and others will not be satisfied or secure. In this sense happiness may be in part a personality trait, which is a cause of behavior as well as an effect. From this point of view, the unhappy person is not unhappy because he was once evicted; rather being unhappy and perhaps angry, he was unable to defer gratification (i.e., save money) and avoid an eviction.

Unlike internal control, however, there has not been enough research with happiness to indicate how much of happiness is a stable personality trait and how much is an end result of the success one has had in living. For the sake of the argument here, we will make the reasonable assumption that it is not predominantly one or the other. This is what common sense would suggest; but being wary of common sense, we also have done some analysis. The very strong associations between happiness and internal control we can take as good evidence that a portion of happiness is a personality trait. For if happiness were entirely ephemeral, why would it be more strongly associated with control than it is with income, education, or being stably married, which are more obvious causes of happiness?

We also have one simple tabulation to show that happiness is not entirely a stable trait. Divorced and widowed respondents are unhappy; but respondents who are married for the second time are not. Since people in their second marriage must have once been widowed or divorced, it follows that they became unhappy when their first marriage ended, and then became happier again when they remarried. To argue that unhappiness causes divorce would imply that unhappy people are also the ones who divorced and remarried; this is contradicted by the data.

[3] Bradburn, *op. cit.*, p. 105. Data given on p. 45 of *The Structure of Psychological Well-Being*.

Summary

Happiness, as measured by self-report, has proven to be a reliable measure in surveys. In our data, happiness is associated with high internal control and with indicators of inhibition, especially in relation to race. In terms of some objective measures of "success," happy people are relatively free from anxiety; they are financially secure, married, satisfied with their jobs, unafraid. Psychological security seems at least as important as financial security to being happy, however. Thus happiness seems to be both means and end: a personality trait as well as a response to the course of life.

8

Personality and Achievement

At this point, let us pause to summarize the four preceding chapters. In each of those chapters we developed a personality trait and studied its impact upon achievement. In this chapter we will draw the personality factors together, and examine more systematically their impact on achievement. So far in this analysis we have used five personality traits. Let us quickly review them.

1. *Internalization of anger.* We have argued that the degree to which one is unable to express anger is an important characteristic—the person who cannot express anger internalizes it in the form of depression or self-deprivation, perhaps finds his anger exploding uncontrollably at the wrong times, and is immobilized from competing with others because of his fear of his own unexpressible anger or his fear of other people's retaliation.

2. *Expression of aggression.* The other side of the aggression coin is the expression of aggression in socially unhealthy ways, through fighting and acts which result in being arrested, for example.

3. *Internal control.* Internal control, we argue, is the degree to which one believes that he can manipulate his environment to get what he wants. The person who believes that his future can be controlled by him (that he has internal control over events) is more likely to plan for success, to make sacrifices in order to obtain it, to be more self-confident and trusting.

4. *Self-esteem.* Self-esteem, as we have measured it, reflects a willingness to see oneself as a special person—someone different from and better than other people in at least a few ways. The person with high self-esteem should, we argue, have more confidence and more ability to compete with others for what he wants.

94

5. *Happiness.* Happiness has two components—a stable and a changing one. The changing component of happiness is the extent to which a person can be pleased or made happy by recent events. The stable component of happiness is a sort of general "good mental health" which probably reflects more than anything else a low level of internal personal anxiety.

We can perhaps add to these five variables a sixth factor which is not a personality trait at all, but an attitude. This is the amount of fear of, and hostility toward, whites—what we have called anti-white feelings.

The Personality Factors

We have discussed the interrelationship of the internal control, self-esteem, and expression of anger variables. In Table 8.1, we present these relationships again, and add the happiness, fight and arrest, and anti-white variables, showing all the interrelationships in a matrix of correlations (although in this case our correlations are shown as gammas rather than the more conventional correlation coefficients). Data are given separately for black men and women, men in the upper right triangle of the matrix, women in the lower left triangle. Examining the matrix we see that for both men and women, the variables can be grouped into two uncorrelated factors. Recalling anger and having high self-esteem comprise one factor which we call *assertiveness,* and internal control, control of aggression, happiness, and low anti-white feelings comprise the other, called *security.* Specifically, respondents who can recall being angry tend to have high self-esteem but they are less likely to avoid fights or arrests, have high internal control, be happy, or have low anti-white feelings. On the other hand, persons who have high internal control tend to fight and be arrested less, to be much happier and have much lower anti-white feelings.

These two clusters seem to be related to two broad dimensions of personality. The man or woman who has high self-esteem and is uninhibited in expressing

TABLE 8.1

Associations between Personality Variables[a]

Variable	Recall anger	High self-esteem	No fight or arrest	High internal control	Very happy	Low anti-white
Recall anger	×	.19	−.29	−.13	−.34	−.13
High self-esteem	.11	×	−.11	.11	−.01	−.04
No fight or arrest	−.15	−.07	×	.21	.23	.18
High internal control	−.12	.12	.15	×	.45	.46
Very happy	−.35	−.04	.27	.35	×	.38
Low anti-white	−.02	.09	.07	.25	.21	×

[a]Men above diagonal, women below.

anger may be thought of as an assertive person—someone who can compete easily with others and work for what he wants out of life. On the other hand, we might think of the person who is happy, tends not to be fearful or hostile around whites, and has high internal control as someone who is relatively free from anxiety—a comfortable person with a high level of emotional security.

The question of why assertiveness and security have a slight negative relationship was discussed briefly in the chapter on anger and in more detail in the chapter on self-esteem. To summarize, it seems to us that the expected type—the person who is high in assertiveness and high in security—does exist. These are both common middle-class traits and many middle-class blacks have them. However, there is a tension between assertiveness and security which reflects itself in the even larger number of people who possess one of the two traits but not the other. For example, one type of apparently secure person who is not assertive is the person who thoroughly internalizes his aggression and is unwilling to express himself, either by saying he is different from other people in any way, unhappy, or that the world may mistreat him in some way. This is perhaps the classical portrait of Uncle Tom, who is convinced that if we will only work hard and not cause any trouble, everyting will work out in the end. At the opposite extreme is the person who externalizes his hostile feelings, blaming the world rather than himself for everything. This defensive posture is one that enables him to say, "I'm fine, but the world—and white people in particular—pick on me and I'm angry about it." We hypothesize that the structure of the black personality differs from that of whites precisely because the black has an overriding need to resolve the issue of whether he should hate white people or not. It is, we think, very difficult to compromise on this issue, to be "moderately anti-white." The tension and anxiety aroused by the race issue press on the subject, forcing him to either "over-internalize" (to become a Tom), or "over-externalize" (to become a militant).

Happiness is highly correlated with low anti-white feeling. The anti-white feelings scale does not measure civil rights militancy; anti-white people are not very active in civil rights work (the gamma between high antiwhite feelings and having been at a civil rights demonstration or rally is .11 for men, .09 for women). Anti-white people are not particularly hostile; they are not more likely to disagree that "most people can be trusted." This means that this anti-white scale does not measure anything analogous to anti-black prejudice among whites. If the anti-white scale does not measure civil rights militancy, prejudice, or hostility, what does it measure? Our assumption is that it measures fear of whites, coupled with a strong sense of anger toward whites. It is plausible that the fear of whites might generate insecurity and produce unhappiness. At the same time, strong feelings of anger toward whites might be psychologically depressing, producing further unhappiness.

The Impact of Personality on Achievement

How important are these personality factors in explaining achievement? Obviously they are not the only factors involved, for we know that educational opportunities, racial discrimination, the state of the economy, and other conditions are important.

Table 8.2, which presents the association between personality factors and achievement, shows a large number of positive associations. The eight achievement variables in Table 8.2 are:

1. Educational attainment.
2. Whether the respondent's first marriage ended in separation or divorce.
3. Response to "Have you ever tried to find a home in a neighborhood which was all white or mostly white?"
4. Home ownership.
5. Score on a scale of financial responsibility, based on having a checking account, savings account, life insurance, stocks and bonds.

Finally, for men only:

6. Number of jobs held in past 5 years.
7. Personal income.
8. Ability to name an employer where respondent could get job now.

In Table 8.2, we have presented the associations separately for northern-born and southern-born respondents, and for men and women, primarily in order to make this table consistent with some we will present in Chapter 10, where sex and region affect the causes of these personality traits in very interesting ways. However in Table 8.2 there are few interesting sex and region differences, so let us summarize in Table 8.3 by simply averaging the four associations to get a single number.

Although these associations do not control for social status, we have added education, as both a column variable—an indicator of achievement—and as a row variable—a cause of achievement. This enables us to determine immediately whether any of the associations between a personality trait and an achievement indicator are spurious (in other words, are simply due to the fact that the personality trait tends to appear only in well-educated people who would achieve because of their education even if they didn't have the trait in question).

The data are reassuring on this score. First, only internal control is highly correlated with education—the other personality traits are not so closely linked with education. Second, education is not usually the best predictor of achievement; in five cases at least one of the personality traits is a better correlate with

TABLE 8.2
Associations of Personality with Personal Success Variables (Gammas)

	High education	Not divorced	Tried for integrated house	Own home	High finance score	Few job changes	High income	Know of job
Security: control of aggression (no fight or arrest)								
Northern men	.35	.27	−.14	.36	.34	.36	.11	−.25
Southern men	.19	.50	.11	.15	.23	.34	.19	.05
Northern women	.31	.40	−.16	.27	.30			
Southern women	.20	.42	.04	.33	.23			
High internal control								
Northern men	.22	.12	−.09	.30	.34	.15	.13	.00
Southern men	.33	.27	−.11	.23	.34	.17	.20	.17
Northern women	.35	.14	.24	.35	.40			
Southern women	.45	.12	−.03	.24	.40			
High happiness								
Northern men	.21	.25	.07	.35	.19	.39	.24	.29
Southern men	.11	.37	−.19	.21	.31	.12	.06	.15
Northern women	.24	.40	−.04	.63	.44			
Southern women	.15	.28	−.12	.30	.29			
Low anti-white feeling								
Northern men	.13	.30	.06	.47	.22	.15	.05	−.13*
Southern men	.14	.01	−.01	.16	.12	.12	.03	.10
Northern women	.20	−.22	.25	.28	.34			
Southern women	.07	−.10	−.13	.12	.10			
Assertiveness: can recall anger								
Northern men	.11	.31	.61	−.05	.08	−.27	−.04	.41
Southern men	.14	.00	.35	−.17	.12	−.26	−.09	.34
Northern women	.10	.06	.32	−.03	.10			
Southern women	.15	−.03	.19	−.03	.12			
High self-esteem								
Northern men	.21	.04	.03	.04	.10	.01	.13	.43
Southern men	.18	−.05	−.05	−.08	.16	−.03	.06	.23
Northern women	.23	−.03	.35	.08	.13			
Southern women	.13	.00	.12	.11	.17			

the achievement variable. The two variables which correlate better with education are the financial responsibility scale and income. In both these cases, however, the associations with education are not much stronger than the associations with internal control. Thus we are on safe ground in concluding that the larger associations in Table 8.3 can not be attributed entirely to social status.

There is a more difficult methodological problem in assessing the effects of personality on achievement. Measures of personality such as internal control or

TABLE 8.3
Personality and Achievement: A Summary of Table 8.2 (Partial associations, controlling on sex and region of birth, γ)[a]

	High education	Not divorced	Tried for integrated house	Own home	High finance score	Few job changes	High income	Know of job
Security								
Control of aggression	.26	.40	-.04	.28	.32	.35	.15	-.10
Internal control	.34	.16	.00	.28	.37	.16	.16	.08
Happiness	.18	.32	-.07	.37	.31	.26	.15	.22
Low anti-white	.18	.00	-.04	.26	.20	.13	.04	-.01
Assertiveness								
Recall anger	.12	.08	.37	-.07	.10	-.26	-.06	.38
High esteem	.19	.02	.11	.04	.14	.00	.10	.33
Education		.15	.20	.18	.47	.07	.22	.29

[a]The association are the averages of the 4 (or 2) associations given in Table 8.2.

happiness are taken on the adult respondent and may be as much a consequence of his economic position as a cause. Generally there is no solution to the problem of causation in this type of situation. We can show that a personality trait is associated with high income; but whether it causes the high income or the income is the cause of the trait cannot be determined in any completely satisfactory way.

In Chapter 5 we attempted some empirical tests and became convinced that internal control is in fact a cause of achievement, rather than a mere effect. In Chapter 6 we discussed the consequences of self-esteem. In Chapter 4 we argued that expression of anger was a cause of behavior, rather than an effect. Unfortunately, in Chapter 7 we were unable to decide how much feelings of happiness were causes of achievement; we are fairly sure that happiness is *both* a cause and an effect, but whether it is more cause than effect, or more effect than cause, we could not determine.

In Table 8.3, we will read the associations first as estimates of the upper limit of the effect of personality; personality cannot have more influence than the gammas indicate, although it might have less influence. We will then draw upon the analyses in the earlier chapters to identify those associations where there is some evidence that the personality factor is the cause of achievement.

Overall, the four factors which show the largest association with achievement are education and the first three variables in the security cluster: internal control, control of aggression, and happiness. For example the association between internal control and the "finance scale," is always above .30, regardless of sex or region, and averages .37.

There are two achievement measures which are not positively associated with the "security" personality cluster; these are whether the respondent knows of a job opportunity for himself and whether he has ever tried to obtain housing in a white neighborhood. With the other six achievement variables, the gammas for control of aggression, internal control, and happiness are over .25 twelve times out of a possible eighteen, and are never lower than .15.

The two measures of achievement which do not correlate with the security group—knowing of a job opportunity and house-hunting in a white area—are also the two measures which correlate positively with the assertiveness cluster. We interpreted this earlier as meaning that while security is a prerequisite for leading a stable goal-oriented life, assertiveness is important in enabling people to strike out in creative ways, or to succeed where one must compete more aggressively.

The association of assertiveness with the other six measures of achievement are mixed. Although expression of anger and self-esteem are both positively associated with education, they also correlate with changing jobs and not owning a home. The associations with income and divorce are mixed. Assertiveness is associated with scoring high on the finance scale but the gammas are all small.

In the preceeding chapters we selected some of the relationships shown in Table 8.3 to examine in detail, and to see if persuasive evidence could be developed to support true causal relationships. In some cases the relationship is one which has been discussed or tested by other writers as well. There seems to be good evidence, either in this study or elsewhere, to interpret some of the associations in Table 8.3 as true cause-and-effect relationships.

First, let us consider the pattern of positive correlations with the "security cluster."

1. *Security and education.* It seems very likely that this association runs in both directions. Being a high school graduate should make one happier, more confident about the future, less trouble-prone and hence less anti-white. But security probably also causes the respondent to obtain more education. The high school student who stays out of trouble, believes that working hard will pay off, is more contented and less hostile, should be less likely to drop out of school.

2. *Security and marital stability.* To some extent, the kind of personality that does not give up on school will also not give up on marriage. However, we are not convinced that is a strong effect. It seems more likely that the divorced person will be unhappy because he or she is lonely, and will get into fights and be arrested more often simply because his or her social life involves dealing with more people and being in public more, and particularly because it involves competing for sexual attention.

3. *Security and homeownership, financial responsibility, and job stability.* In the chapter on internal control, we concluded that internal control could be read as a cause of homeownership, financial stability, or job stability. It seems likely that control of aggression can also be read as a cause; the person who has no serious aggression problems has less difficulty dealing with his employer or in saving money and eventually should be more likely to own his home. We are less willing to conclude that happiness causes these forms of behavior; being a homeowner or having money in the bank may make the respondent happy.

4. *Security and income.* In Chapter 4 we argued that excessive job changing was probably one reason why people with aggression problems earned less money. It also makes sense that the person with a sense of internal control should have higher income—he might work longer hours, for example.

5. *Security and knowing of another job.* If we look at Table 8.2, we see that the security variables are associated with knowing of a job opportunity only for Southerners; for northern-born respondents the association is zero or negative. In our discussions of aggression and internal control, we argued that excessive inhibition of aggression, and a high internal control score based on an inability to become angry, are inhibiting, and prevent the respondent from competing in risky situations. Job-hunting is one such example; trying to escape from the ghetto is another.

6. *Assertiveness, integration, and job knowledge.* In Chapter 4, we argued that the uninhibited person can compete more easily and thus that assertiveness is a trait which causes achievement in competitive situations. Thus the assertive person

is more willing to move into a white neighborhood, more optimistic and knowledgeable about job opportunities.

However the data clearly indicate that, on balance, expression of aggression and believing that one is above average are not useful for blacks. Both are associated (slightly) with being better educated; but they are generally dysfunctional traits. Self-esteem is very slightly associated with most of the measures of achievement, but this is probably only because high-esteem subjects are better educated. High self-esteem may be primarily a matter of reaction formation—empty boasting to cover a sense of insecurity.

In this analysis we were quite surprized that self-esteem and expression of anger were not valuable personality traits. Of course, they may be more valuable for whites, but we wonder if we wouldn't be surprised there as well.

One reason why we expected aggression and self-esteem to be more positively related to achievement is that we are caught up in the tenor of the times: the 20th century is a time of pressing for greater freedom and self-expression. Traditionalism has been attacked by science, atheism, cultural relativism, the sexual revolution, and free enterprise; the most important impact of psychoanalysis has been to preach that one should be free to express oneself, to overcome inhibitions. As scientists we should not be so strongly prejudiced by the new conventional wisdom, and should not have been surprised that old-fashioned virtues are still rewarding.

Summary

In general the personality traits developed in Chapters 4-8 fall into two clusters: a "security" group consisting of internal control, happiness, avoiding fights or being arrested, and a low level of anti-white feelings. This group is positively correlated with most measures of achievement, including higher levels of education and income, homeownership, financial responsibility, and marital and job stability. However two measures which suggest being able to compete—trying for integrated housing and knowing about a job opportunity—are negatively associated with security.

The second personality factor, which we have called "assertiveness," consists of the ability to express anger and high self-esteem. It is negatively correlated with the security items and its correlations with the achievement variables are almost exactly opposite to those of the security group.

9
The Broken Home

So far we have argued that in analyzing black achievement one should be concerned with the personality differences between blacks and whites. In the next three chapters we will focus on two of the factors which are special in the black experience to see if these factors can in any way account for the black-white differences we have observed. Blacks have traditionally had less stable families than whites; we shall examine both the causes and the consequences of marital instability in this chapter. Then in Chapters 10 and 11 we will look at the psychological impact of discrimination and segregation.

Census data show that families headed by women—including widowed and single women as well as those who are temporarily or permanently separated from their husbands—are more frequent among the black population than the white. In March 1971, 28.9% of all nonwhite families were headed by women, whereas only 9.4% of white families were of this type.[1] The figure for blacks represents an increase of about 60% since 1960. While the proportion of female-headed families in both the white and black populations is greatest at the poverty level (i.e., family income below $3000) and steadily decreases as family incomes increase, the proportion for blacks remains about twice as great as that for whites at every income level.[2] Thus female-headed families are not only a consequence and cause of poverty (since women earn less than men and may not work if they have dependent children) but are also related to race. While numerous researchers,

[1] U.S. Department of Commerce, Bureau of the Census (1971), *The Social and Economic Status of Negroes in the United States, March 1970*, p. 107, U.S. Government Printing Office, Washington.
[2] *Ibid.*, p. 109.

including Moynihan, have been concerned with this fact, no studies of a representative adult sample have attempted to trace the causes of marital instability among blacks.

In this study both white and black respondents were asked, "Did you always live with both your mother and father until you were 16 years old?" Forty-three percent of the black respondents said no, compared to only 20% of the white comparison group.[3] But this actually understates the differences in the stability of white and black families. When the respondents were asked why their home was broken (the responses are given in Table 9.1) blacks were twice as likely to have lost a parent through death, and five times as likely to have their parents separate or divorce.

Before going further there is a technical problem which must be considered. At the time our respondents were born approximately 2% of all white births and 20% of all black births were classified as illegitimate. Illegitimacy was not a category on the card given to the respondents when they were asked how their family broke up. How did the illegitimate children respond to the question? The

TABLE 9.1
Respondents' Description of Their Family Situation Until They Were 16[a]

	White(%)	Black(%)
Lived with both parents	80	57
Cause of breakup:		
Father died	8	12
Mother died	5	9
All deaths	13	21
Father deserted	1	3
Mother deserted	0	1
Parents separated	1	14
Parents divorced	3	5
All voluntary breaks	5	23
Respondent left home	1	2
Other	2	2
	101	105
N	(345)	(4153)

[a]Due to differences in wording, blacks were permitted to give more than one answer, but whites were not.

[3] Census data show that in 1966 (when this survey was taken), 29% of all nonwhite children of family heads were *not* living with both parents; 9% white children were *not* living with both parents. Since the survey question asked respondents about their first 16 years of life, we would expect the proportion of respondents who report *not* living with both parents to be greater than the census figures for a single year.

one possibility is that the respondent would say his or her father died. In fact, the number of respondents who listed their fathers as dead is slightly less than what one would predict based on the death rate for black men during this period. (In 1940 the death rate for black men in their thirties was over twice that for white men.) Respondents were also asked at what age their family broke up, and using this information we find that there is indeed a larger-than-predicted number of respondents whose fathers died before they were one year old. These cases—about 1% of the sample—may be illegitimate children who were told that their fathers died.

Our best guess is that illegitimacy does not create a technical problem in our analysis because most technically illegitimate children did in fact know their real fathers.[4] (And if not, they may still have been raised in a two-parent home.) If their natural parents remained living together, or married soon after the child was born, the respondents would consider themselves as coming from a stable home. If their parents later broke up, the respondents would then say that their parents separated, and this would appear as a broken home in our analysis. Thus we will ignore the issue of illegitimacy in this chapter.

Returning to Table 9.1 we see that for both whites and blacks, more fathers than mothers died. This is consistent with the higher death rate of men in this age group. For whites the small number of voluntary breakups are concentrated in the divorced category, while for blacks the most common response is that the parents were separated. In this analysis we will not distinguish between families broken by divorce and those broken by separation or desertion. There may be important legal differences here, but the social process of the breakup is similar and the consequences for the children should be roughly the same.

Table 9.2 shows the composition of the broken homes in the sample. Approximately half the respondents from broken homes lived with their mothers, who generally did not remarry, and rarely were there male relatives at home. Thus over half of the respondents who lived in broken homes lived in all-female households, and only one-fourth grew up in households with a married couple. This means that 68% of the *entire* sample grew up in households with a married couple.[5]

---

[4] Studies of lower-class black family life show that paternity is usually acknowledged and is even a source of pride for the man. Also the man often continues to see his child and its mother. See Lee Rainwater, *Behind Ghetto Walls* (1970), Aldine, Chicago. Joyce Ladner (1971), *Tomorrow's Tomorrow: The Black Women,* Doubleday, New York.

[5] In his study of stable middle-class black marriages, Scanzoni classifies families of origin with a step-parent as "stable," thus raising to 73% the proportion of his respondents coming from stable homes. (58% of his sample lived with "natural" parents during most of adolescence; in our sample, 57% lived with natural parents, but only an additional 6% with a parent and step-parent.) This is pointed out to illustrate that studies are inconsistent in their definitions of "broken" homes. See John H. Scanzoni (1971), *The Black Family in Modern Society,* pp. 43-44, Allyn and Bacon, Boston.

TABLE 9.2
Those Respondents Lived with after Breakup

Lived with:	Percentage
Women only:	
Mother only	36
Mother and female relatives	8
Female relatives (grandmother, aunt)	11
	55
Men only:	
Father only	7
Male relatives (uncle, grandfather)	1
	8
One parent plus relatives of opposite sex:	
Mother with relatives (including one or more men)	2
Father with relatives (including one or more women)	1
	3
A couple:	
Grandparents	5
Aunt and uncle	6
Mother and step-father	11
Father and step-mother	4
	26
Lived alternately with each parent	2
Child lived alone	2
All other	2
No answer	2
	8
	100 ($N = 1765$)

The Effects of the Broken Home: Education

A good deal has been written about the effects of the broken home, but the data have been meager and the results sometimes inconsistent. The data we have is unique in that it is a large random sample of adults from broken and stable homes, so we can measure the long-term effects of family stability with reasonably reliable statistics. Thus we will have answers to some previously unanswerable questions. To dispose of some "null" findings first, there seems to be little relationship between coming from a broken home and level of education for men or verbal test score for either men or women. The gammas between family stability and verbal achievement are .08 for men, .03 for women; respondents from broken homes have slightly lower verbal achievement than those from stable homes, but the differences are negligible.

There is a definite relationship between family stability and education for women: 52% of the women from stable homes finished high school, and 21% attended one or more years of college. For women from broken homes, the comparable figures are 40 and 12%.[6] This seems to be because girls from broken homes are more likely to marry early. Thirty-two percent of the married female respondents who came from broken homes were married before they were 18 compared to only 24% of the women from stable homes. Most of these early marriages would result in the women dropping out of high school. In the case of men, we see almost no difference at all in educational attainment; slightly less than half of the men from stable homes finish high school and the same is true for those from broken homes.

Actually the pattern is a little more complex than this. Approximately two-thirds of our respondents are migrants from the South, and Southerners are much less likely to have finished high school than respondents raised in the North. But if the family breaks up, this provides a convenient opportunity for the mother and children to move to the North, and therefore the southern-born boy from a broken home is more likely to have the opportunity of a northern education. For example, 19% of southern-born respondents from stable homes moved north before they were 10 years old, but 22% of respondents from broken homes moved this early in life. Only 31% of the respondents from stable homes who stayed in the South past the age of 10 finished high school. Only a small number of these migrated north to complete high school in a northern city. In contrast, 36% of the respondents from unstable homes finished high school, apparently because they moved north as teenagers and were able to finish high school in a northern city. Thus the complete story is that coming from a broken home has both positive and negative effects on educational attainment. Its positive effect lies in the fact that Southerners whose families break up have more opportunity to complete high school by coming to the North. The definite negative effect is among respondents who were already living in the North when their families broke up. Sixty-five percent of the respondents from stable homes who were living in the North at age 10 finished high school, compared to only 57% of respondents from broken homes.

---

[6] This result is surprising since Moynihan and others have emphasized negative effects of the broken home for males' educational achievement and implied that effects for females were *not* negative.

In any case, none of these differences for men is impressively large. This agrees with Elizabeth Herzog and Cecelia Sudia, who review the literature in *Boys in Fatherless Families*.[7] They write:

> With regard to academic performance it seems unlikely that father absence in itself would show significant relation to poor school achievement, if relevant variables (including type of father absence) were adequately controlled.

The Effects of the Broken Home: Income

In Table 9.3 we look at the relationship between stability and income of men, and the story is quite different. Men from broken homes, whether broken by death or separation, have considerably lower incomes. In the case of respondents with separated parents, incomes are nearly $800 lower than those of respondents from stable homes.

While this finding may not surprise the nonsociologist who has always assumed that broken homes were unfortunate, it is quite surprising to someone familiar with the research literature. Studies of the long-term income effects of the broken home have not been done before, and there is nothing in the literature to suggest that there should be this sharp difference. .

We have already seen that the differences in level of educational attainment or verbal test scores are too slight to explain this sizable difference in income. Nor can these differences be attributed to region of birth, age, or mother's education.

TABLE 9.3
Income of Men by Stability of Home

Income ($)	Lived with both parents until age 16 (%)	One or both parents died (%)	Parents separated or divorced (%)
0-3499	13	18	23
3500-5499	31	32	38
5500-7499	34	37	29
7500+	21	13	10
	99	100	100
N	(1047)	(283)	(341)
Median	$5827	$5477	$4944

[7] Elizabeth Herzog and Cecelia Sudia (1970), *Boys in Fatherless Families*, U.S. Department of Health, Education, and Welfare, Office of Child Development, Childrens Bureau, Washington, D. C.

It is difficult to control adequately on parental socioeconomic status since the father's occupation is irrelevant to families where the father is absent. One measure of family status which is applicable (and known to the respondent) in almost all families is the education of the mother. The mother's education is generally a better predictor of the children's behavior than is the father's education. However it would be impossible to explain the negative effects of coming from a broken home by arguing that children from stable homes have better educated mothers, since actually the mothers in stable homes are slightly *less* likely to be high school graduates than the mothers in broken homes.

This whole question of whether family stability is the factor which explains these lower incomes for men or whether there is some other background characteristic of more importance was checked with a series of regression equations which included the education of both parents and the respondent's own education. In these equations stability[8] was found to have an effect on income of $500 per year, independent of parents' education and the respondent's own education. For comparison, the effect of having a father who graduated from high school is only an increase of $200 per year. (Mother's education also has an effect of about that size.) These regression equations estimate that the high school dropout earns $360 per year less than the high school graduate and the man with some college training makes only $410 more than the man with a high school diploma. This means that coming from a stable home is an economic resource equivalent to about two or three years of formal education, and is nearly three times as valuable as having a father who was a high school graduate.

Possible Explanations: Poverty

Using mother's education as a control variable is an inadequate test of one obvious explanation for the broken home effect. Broken homes are often poor, even when the mother is well educated, and it might be that boys from broken homes have lower incomes because they have not been able to compensate for the limited resources of their poor upbringing. This explanation seems incorrect for three reasons:

1. If the poverty of the broken home is critical, then the greatest effect should occur in families where the main earner, the father, died. In the case of separation there is at least a possibility that the father will continue to contribute income. But when we look at respondents' incomes, we see that men whose

---

[8] Note that these equations refer to the effect on income of family instability *per se* and make no reference to the *cause* of instability (i.e., divorce, separation, desertion, or death) or the *type* of broken home (i.e., male- or female-dominant). Hence these findings should *not* be interpreted as the effects of father absence.

fathers died have unexpectedly high incomes: a median of $5642, which is only $185 below that of men from stable homes. The same pattern can be seen another way. We asked the respondent whom he lived with after the family broke up. If poverty were the right explanation, then those respondents who lived with their fathers or other male relatives should have had, on the average, less economic hardship. But when we look at the incomes we see that those men who lived with their fathers after the family had broken up had the *lowest* incomes. The median incomes for these respondents is $4639 (based on 121 weighted cases). This is considerably below the median income of respondents who lived with their mothers or other female relatives.

2. A second reason why we cannot attribute the effects of broken homes to poverty is that provetty does not appear as that powerful a factor. For example, if living in poverty had a major effect on the respondent, how can we explain the fact that growing up in the South does not reduce income? Certainly poverty is more severe there. Secondly, parents' education is not a very strong predictor of respondent's income. Certainly the stable home where the mother or father is not a high school graduate will be on the average somewhat poorer than the home where either parent was a high school graduate, but this does not have a large effect on the future earnings of the children.

3. If childhood poverty were the major factor in explaining the low incomes of respondents from broken homes, we would expect, first, that children from broken homes would have less education, and secondly, that those who manage against these odds to obtain a better education would escape most of the unfortunate consequences of their childhood. But the effects of family stability on education are not very large for men. More important, the effects of coming from a broken home are actually larger for respondents who graduated from high school than for those who did not graduate. Men who did not finish high school earn a median of $5344 if they grew up in a stable home, and $5026 if their home was broken—a loss of $318. High school graduates from stable homes earn $6187, $800 more than the $5387 earned by graduates from broken homes. The respondents who have managed to graduate from high school show the greatest income loss because of their broken family. The same pattern appears if we use high school grades or verbal test score as a variable. The high-achievement students, those with good grades and good test scores, show the greater loss of income due to family instability (data not shown).

It seems to us that these three points present a good argument that there is something other than poverty which explains the bad effects resulting from broken homes.

The Female-Dominant Home

Let us next present some negative evidence regarding another common hypothesis: that the trouble with black families is that the women are too powerful, the men too weak.

From the analysis we have done so far it does appear that black men suffer much more than black women from discrimination and show more psychic damage as a result. As we shall see, it looks as if men create more marital problems than women. But writers frequently add a corollary to this: that even in stable homes the emasculated father provides a poor role model for black males. The weak father is not much better than the absent one. To test this line of argument we asked the respondent four questions about his family (see Table 9.4). In each case he was asked to say which member of the family had the most power or influence in some particular area. If the respondent named a woman (his mother, stepmother, aunt, etc.), this was counted as an incident of "female dominance." There is a consistent tendency for blacks to name females more than whites do in response to these questions. However the differences are not very large. Partly they can be explained by the fact that blacks were more likely to live in fatherless homes. When allowance is made for this, the white-black differences are smaller. It is also worth noting that on three of the four questions

TABLE 9.4

Measures of Dominance of Male and Female Adults in the Family for Whites and Blacks

	White		Black	
	% giving male figure[a]	% giving female figure[a]	% giving male figure	% giving female figure
When you did something wrong, who usually punished you?	26	56	20	68
Who kept track of the family money?	35	53	27	61
Thinking back now, who do you think made the important decisions in your house when you were a child?	43	39	35	52
Who did you admire or respect the most?	27	39	19	56
	(N = 346)		(N = 4170)	

[a]Rows, within each race; do not add up to 100% because many respondents answered "don't know" or "both."

whites name female figures more than male figures. Apparently the white family should also be viewed as a female-dominant one.

How then should we interpret Table 9.4? One possibility is that the whole notion of the black family as more matrifocal may be in error.[9] It is true that there has not been very much good research on this whole question. The hypothesis that black men should appear as much weaker figures in the family is primarily deduced from a general conception of black men as having difficulty establishing and maintaining masculine status in the face of discrimination.[10] There is not very much research on this, but the whole argument seems plausible and nothing else in our analysis causes us to question it. Merely the fact that a man has higher status than a woman, however, does not necessarily mean that he will exercise more influence or take on a larger portion of child-rearing activities, although it may mean that his children will admire or emulate him more. The man who has higher status than his wife may use that status to push the child-rearing activities onto her shoulders.[11]

The next question is "does coming from a broken home lower one's income because the home was dominated by females?" If so, we should find that men from stable female-dominated homes should also have relatively low incomes compared to men from stable homes where fathers were dominant. But as Table 9.5 indicates, the results are the opposite. The stable female-dominated

TABLE 9.5
Effect of Sex Dominance and Stability of Home on Income of Black Men

	Broken home	Stable home
Strong female dominance	$5459 (354)	$6083 (229)
Female dominance	$5143 (72)	$6071 (196)
Slight female dominance	$4588 (79)	$5664 (318)
Neutral or slight male dominance	$5714 (23)	$5647 (161)
Male dominance	$5125 (87)	$5800 (85)

[9] For a discussion of this, see Herbert Hyman and John Skelton Reed (Fall 1969), "Black Matriarchy Reconsidered: Evidence from Secondary Analysis of Sample Surveys," *Public Opinion Quarterly, 33,* 346-354.

[10] Pettigrew discusses the research on the sex-role adoption of males raised in fatherless homes, but the evidence he presents is hardly conclusive and certainly need not apply to males raised in stable, female-dominant homes. Studies cited are based on small samples from widely divergent populations (not all of them black) in this and other cultures. See Thomas Pettigrew (1964), *A Profile of the Negro America,* pp. 17-21, Van Nostrand, Princeton.

[11] The novels which present a patriarchal family often show the father as a very aloof man who is rather inactive in the home.

home produces slightly higher incomes than the stable male-dominated home. This means that the fact that the broken home is dominated more often by a women should work to the economic advantage of the child from the broken home rather than to his disadvantage. In any case, the differences between broken and stable homes remain very large when sex dominance is controlled in Table 9.5.

Apparently having no father and having a relatively weak father (as we have defined weakness) are completely different things. The two-parent home where the mother makes the decisions may not produce "masculine" boys (we don't know), but it does produce economically productive ones. Is this plausible? It may well be that much as some might cherish masculinity, it is an irrelevant or even harmful trait for industrialized society. Perhaps masculinity was a useful force to put behind the plow that broke the plains, but the traits which work in today's society (such as dependability, thoroughness, trustworthiness, deference to authority), are as much or more "feminine" than "masculine."

We hope someone else will direct research efforts to this whole topic of matrifocal families, masculine self-identification, and income. We suspect it is a complicated topic worth the effort of further research.

Broken Homes and Personality

Having found no evidence that poverty or female dominance is the factor which makes the broken home so unfortunate, let us consider possible personality variables. Table 9.6 shows that coming from a broken home is strongly associated with the six personality variables we have used in the previous four chapters. The men from broken homes are much more likely to be in fights or arrested, have considerably lower sense of internal control, and are much less happy. But they are more able to express anger and their self-esteem is not depressed; the respondent from the broken home is thus very low on the security dimension and slightly above average in assertiveness. The effects in Table 9.6 are generally much larger for men than for women. This is consistent with the widely held belief that male children suffer psychologically more than females from a family breakup.

The pattern that we see here—of men from broken homes being pessimistic, unhappy, and angry—is consistent with the fact that much research has shown that fatherless boys are more likely to be delinquent. Herzog and Sudia, in their review of the literature, find that studies of delinquency have consistently shown broken homes to be a relevant factor. It is also consistent with Bacon, Child, and Barry's analysis of cultural data which shows that those societies in which father absence is the normal rule in child-rearing have higher rates of crime:

The cross-cultural findings indicate that a high frequency of both theft and personal crime tends to occur in societies where the typical family for the society as a whole creates lack or limitation of opportunity for the young boy to form an identification with his father.[12]

TABLE 9.6
Attitudes and Behavior by Stability of Home

	Lived with both parents (%)	One parent died (%)	Parents separated or divorced (%)	Difference (γ)	
				Effect of death	Effect of separation or divorce
MEN					
High internal control	62	46	51	−.31	−.22
Consider self "happy" or "pretty happy"	80	72	75	−.22	−.14
Have not been in a fight as an adult	69	52	54	−.35	−.31
Have not been arrested	68	48	59	−.39	−.19
Can recall last time angry	58	65	70	+.15	+.26
High self-esteem	48	47	45	−.02	−.06
N	(1120)	(300)	(379)		
WOMEN					
High internal control	53	46	50	−.14	−.06
Consider self "happy" or "pretty happy"	81	76	77	−.15	−.12
Have not been in a fight as an adult	81	80	75	−.03	−.17
Have not been arrested	93	91	87	−.14	−.33
Can recall last time angry	69	73	76	+.10	+.17
High self-esteem	38	34	39	−.09	+.02
N	(1234)	(421)	(512)		

[12] Margaret K. Bacon, Irvin L. Child, and Herbert Barry, III (1963), "A Cross-Cultural Study of Correlates of Crime," *J. Abnorm. Soc. Psych., 66,* 291-300. Reprinted in Ivan D. Stiener and Martin Fishbein (1965), *Community Studies in Social Psychology,* Holt, Rinehart, and Winston, New York.

The most surprising finding in Table 9.6, however, is that in seven of the eight cases the unfortunate effects of the broken home on the security variables are actually more severe if the home is broken by *death* than if the home is broken by *separation*.

There are three different factors involved in analyzing the impact of voluntary family breakup. First we should consider the impact on the children of the unhappy marital relationship before the breakup. Second is the impact of community disapproval of divorce, the unhappiness of a mother who has been deserted, and the effect this has on children. Third is the impact of the actual loss of one of the parents. Thus one could react to the fact that men with divorced parents have lower incomes by saying that it is not the divorce that hurt, but rather it is the effect of the unhappy parent relationship before the divorce plus the impact of later living with a mother who in many cases has individual personality problems. Many writers today take the position that given an unhappy marriage, divorce may be better for the children than the parents' remaining together. What they are saying is that the actual absence of a parent is less important than the kind of relationship the child had to the parent or parents he lives with. Herzog and Sudia state this case well by saying:

> It would probably be undesirable as well as impossible to restore all absent fathers to the homes they have left. To judge by the research evidence it would also be undesirable to prevent all family breakdown since the evidence indicates that discord and conflict in the home can be more detrimental than father absence; one is forced to prefer a "good" one parent home for a child. Marital counseling may help to preserve harmony and two parents within a home but in some instances divorce may be the more constructive solution for all concerned.[13]

They go on to argue that since there is little society can do to reduce divorce rates, ". . . it may be more feasible to reduce some unfounded anxieties of middle class, one-parent mothers and to offer them such community supports as can be developed than to reduce the divorce rate."

But this argument does not help explain why the man who lost a parent through death should have a depressed income, problems handling aggression, and lower sense of internal control. Death strikes the emotionally strong as well as the emotionally weak. There is no reason to suppose that marriages broken by death are worse than others or that the surviving parent is any more neurotic than anyone else. It must therefore be that parent absence in itself has a more important impact than either the type of parent or the type of marital relationship. For parent absence without these other negative factors (in the form of parent

---

[13]E. Herzog and C. Sudia, *op. cit.,* pp. 97-98.

death) has as much impact on our personality measures as separation through divorce, where these other factors are present.[14]

Broken Homes, Personalities, and Incomes

The next task before us is to determine to what extent the presence of these negative personality factors and aggressive behavior can explain the low income of men from broken homes. Table 9.7 demonstrates that if respondents from stable homes and broken are matched on their arrest record, part of the difference in income disappears. If we look only at persons with arrest records, the difference between those from broken homes and stable homes drops to $209. For those who say they have not been arrested, a large difference of $798 remains.

If the higher arrest rate of men from broken homes explains part of their lower incomes, then we should find (Table 9.7) that the average difference between men from stable or broken homes, after matching them on arrest records, is reduced. The relatively small number of men from stable homes who have been arrested should look like the relatively large fraction of men from broken homes who have also been arrested. One way to estimate the amount of difference remaining after arrest differences have been removed is to compute an average of the differences remaining between stable and broken homes. The differences given in Table 9.7 are $209 among people who have been arrested and $798 among those who have not been arrested.[15] A weighted average of

[14] The possible over-reporting by respondents of the death of parents was noted at the beginning of this chapter.

[15] There is a consistent tendency for the differences between two low-income groups (in this case, the two groups who have been arrested) to be smaller than the parallel difference when two higher-income groups are compared. This is because of the tendency of low income groups not to fall below a "floor," which for men is about $4800 (1965 income) and for women about $2700. This is shown in the following table, using a continuous variable, verbal test score, to divide the population into small income classes:

Verbal Test Score and Personal Income

	Median income ($)	
Test score	Men	Employed women
8–9 (high)	$6857 (119)	$5816 (76)
7	6184 (242)	4600 (104)
6	5644 (431)	3442 (253)
5	5255 (374)	3140 (208)
4	5586 (262)	2783 (113)
3	5037 (165)	2850 (68)
1–2 (low)	4833 (106)	2714 (48)

Note that in the low income range the effect of a decrease in the test score has only a slight effect on income as the figures fall closer to the income floor. This is at least a partial explanation of why the broken home reduces income among arrested men by only $209.

TABLE 9.7
Effect of Arrest Record and Stability of Home on Men's Income

	Arrest	No arrest
Stable home	$5254 (313)	$6065 (729)
Broken home	$5045 (313)	$5267 (374)
Difference	$209	$798
Net Effect of Stability		$577

these two numbers is $577; this can be considered the income loss attributable to coming from a broken home after arrest differences are removed.[16] Since the actual difference between stable and broken homes is $655 before arrest is controlled, this method of handling the statistics allocates $78 as the reduction in income which broken homes cause by increasing the arrest rate, while the remaining $577 is presumably attributable to other factors. However the arrest record has a large response error and consequently the statistical procedure produces too small an effect due to arrest because of this error. Therefore we can conclude that the higher arrest rate of persons from broken homes explains more than $78 of the $655 loss associated with coming from a broken home. A possible counterargument is that the low income of people from broken homes makes them more likely to be arrested. This turns out not to explain the data; the arrest rate for men from broken homes is too large to be accounted for in this fashion.

The higher arrest rate of men from broken homes is only part of a general pattern of instability in their lives. For example, men from stable homes are more likely to own their own homes: for men from stable homes, 13% under the age of 30 and 32% over the age of 30 own homes, compared to 10 and 28%, respectively, for men from broken homes. Table 9.8 shows several other ways in which men from stable homes show more stability and commitment to work. Men from stable homes change jobs less, are more likely to have a secure job position now, work longer hours, and are less likely to have ever been divorced. Each of these factors contributes to their income. Men who have held only one job in the last 5 years earn $865 more than men who have changed jobs; men who work over 48 hours earn $526 more than men who work fewer hours; men who have been divorced or separated earn $375 less than men who have never been divorced.

If we follow the same statistical procedure that we did with Table 9.7, we find that the net difference still attributable to coming from a broken home,

---

[16] See Appendix 5 for a discussion of the measures of partial association used here and elsewhere in this analysis.

TABLE 9.8
Broken Homes and Unstable Occupational Histories of Men

	Stable home (%)	Broken home(%)
Held only one job in past 5 years	50	37
Believe they will not be laid off in present job	84	73
Work over 48 hours	12	9
Have never been divorced (of those ever married)	77	68
N	(1133)	(750)

after number of jobs held is controlled, is only $395—a reduction of $260 from the gross difference of $655. Similarly the effect still attributed to family stability after controlling number of hours worked is $474, a reduction of $181. Finally, if we compare men from broken and stable homes, matching them on their marital history, we find that the income difference due to coming from a broken home is reduced $58. These three calculations suggest that a large fraction of the income loss which the man from a broken home suffers is due to the instability in his personal and occupational history.[17]

We cannot state precisely how much of the effect of broken homes on income can be attributed to this kind of instability other than to say that it seems to be a large fraction. In order to arrive at a precise estimate we would need a controlled experiment which would demonstrate conclusively that job change is a cause of low income more than it is an effect. It is conceivable that some of the job-changing of men from broken homes is an effect of low income rather than a cause. We know that the amount of job-changing is too great to be entirely explained by poverty, just as in Chapter 4, we saw that there was too much job-changing by men who had been arrested to be explained by low income.

Thus we see a pattern—not a complete explanation, but a coherent one. Men from broken homes are more fatalistic, less likely to defer gratification, more likely to express anger or aggression. They are more likely to change jobs, change wives, not buy a house, work fewer hours. All of these factors tend to reduce income. This pattern is consistent with some of the psychological theory regarding the effects of deprivation of parental love (in this case because of the absence

---

[17] These estimates of income explained by arrest rate, number of jobs, hours worked, and marital stability are not independent estimates, and hence cannot be added to produce a single dollar value.

of a parent). Supposedly the child who is deprived of sufficient parental love tends to grow up with a life-long excessive need for gratification and a sense of anger at this deprivation. Thus he is more prone to quit a job because the pay is not enough to make him happy, to get a divorce because his wife does not love him enough, to get into a fight because he has been insulted. This theory is consistent with what we have found, but we cannot test it further with our data.

We can summarize the effects of family stability on achievement by tabulating it against the eight measures of achievement presented in Chapter 8. Table 9.9 shows the result. Family stability is associated with the six types of achievement which are correlated with security, and negatively correlated with job knowledge and trying for a house in an integrated neighborhood. Thus family stability has an effect very similar to internal control, although not usually as strong. Many of the effects are as strong for women as men, but there are some exceptions, and one of them is very interesting. For southern-born women (and 70% of our women respondents are southern-born), there is no association between family stability and the stability of their own marriage.

The Causes of Broken Homes

There are two main approaches to the question "Why are black marriages so unstable?" On the one hand, we can consider it a difference in culture; on the other hand, a difference in social situation. The culture argument would maintain that the matrifocal family consisting of mother and children, with the father only loosely attached, is the accepted family structure in many societies,

TABLE 9.9
Effects of Coming from a Stable Home on Achievement Variables by Sex and Region of Birth (γ)

	High education	Not divorced	Tried for integrated house	Own home	High finance score	Few job changes	High income	Know of job
Northern men	.08	.29	−.24	−.07	.14	.08	.21	−.27
Southern men	−.04	.18	−.02	.21	.17	.20	.22	−.08
Northern women	.40	.29	.08	.19	.25			
Southern women	.14	.04	.14	.03	.02			
Average effect	.14	.20	−.01	.09	.14	.14	.22	−.18

including the lower-class American black subculture.[18] The alternative argument is that black marriages break up, not because the black community endorses divorce, but because the social psychological and economic pressures are so great.

Our data will permit us to reject the cultural explanation, and after doing this we will attempt to unravel some situational factors which cause marital instability.

The Culture of Matriarchy

Journalists have a standard portrait of the three-generation welfare family, consisting of a grandmother (with no husband), a daughter (raised on welfare and also unwed), and her newly born child. It is a compelling portrait, which is intended to suggest that women raised on welfare are themselves less likely to have stable marriages, or more likely to be unmarried mothers.

But data on family of origin is not available in the census, and no one has ever been able to observe the extent to which fatherless daughters are likely to have fatherless children. Our data cannot treat the question of unwed mothers. Only 5% of our female respondents have never married by age 35, and one would not expect a presently married woman to admit that one of her children was born out of wedlock. But our data will tell us whether fatherless daughters are more likely to be divorced or separated, and the answer, given in Table 9.10, is not at all as expected. Forty percent of women from homes broken by divorce or separation had their own first marriages break up; but 36% of women from stable homes were divorced or separated from their first husbands. The difference of only 4% is not worth paying attention to. In the top line of the table, we look only at the responses of presently married women to the question, "Since you have been married, do you think you have had your best times when you were doing things with your spouse or when you were with friends, but without your spouse?" Responses are presumably indicators of their satisfaction with their marriages. Here we see the expected patterns: daughters of divorced parents are less happily married.

Does this mean that the idea of the transmission of marital instability from mother to daughter is a myth? The answer would seem to be positive. While the broken home is apparently poor preparation for a happy marriage, daughters from broken homes are not freer in their initiation of divorce. When we turn to

---

[18]Two schools of thought prevail on this point. E. Franklin Frazier and Moynihan argue that the experience of slavery and economic conditions since emancipation are responsible for the matrifocal family tradition among blacks. Herskovitz argues that the African cultural heritage is responsible. See Melville J. Herskovitz (1941), *The Myth of the Negro Past*, pp. 167-186, Harper and Row, New York.

TABLE 9.10
Attitudes and Behavior Toward Marriage

	Lived with both parents until age 16	One or both parents died	Parents separated or divorced
WOMEN			
Had "best times" without spouse[a] (for currently married), %	13 (740)	20 (271)	23 (278)
First marriages ending in separation or divorce (of all ever married), %	36 (1084)	45 (409)	40 (449)
MEN			
Had "best times" without spouse[a] (for currently married), %	14 (767)	13 (171)	20 (218)
First marriages ending in separation or divorce (of all ever married), %	24 (929)	40 (233)	27 (293)

[a]"Since you have been married, do you think you have had your best times when you were doing things with your husband (wife) or when you were with friends but without your husband (wife)?"

the center column of Table 9.10, the story becomes more complex. We see that women from homes broken by death do have high divorce and separation rates, and those who are married are almost as unhappy as daughters of divorced parents. This agrees with what we have seen in earlier tables; homes broken by death are at least as damaging as those broken by separation.

The divorce rate for women with divorced parents is low, despite their unhappy marriages, because rather than learning that divorce is permissible, they are more determined to keep their marriages together. They are determined not to follow in their mother's footsteps; they are "once bitten, twice shy." Herzog and Sudia cite a study by Judson Landis[19] indicating that "children of [broken] homes are interested in marriage, and determined to make it a success." This interpretation would explain why the divorce rate for women from homes broken by death is high. For these women, like all women from broken homes, are ill-equipped for marriage, but not having experienced their parents' divorce, they are less fearful of this solution to their own marital difficulties.

[19] Herzog and Sudia, *op. cit.*, pp. 56–57; Judson Landis (Feb. 1960), "The Trauma of Children when Parents Divorce," *Marriage and Family Living, 22* (No. 1), 7–13.

The data for men are consistent with this. The third line of Table 9.10 shows that men with divorced parents are more likely to be unhappily married. It does not show a higher rate of marital unhappiness for men from homes broken by death; this is the one uninterpretable inconsistency in the table. In the last line of Table 9.10, we see that the pattern for men is similar to that for women, but more pronounced. Men from homes broken by death are much more likely to be divorced; the 40% divorce rate here is nearly double the 24% rate for men from stable homes. However men from homes broken by divorce do not have a high breakup rate, even though they are more often unhappily married. This again supports our view that children from broken homes are more anxious to avoid divorce. The effects of broken homes are stronger for men, suggesting that divorces are more often initiated by men. As we shall see, there is considerable evidence in our data to support this hypothesis.

More important to the cultural hypothesis, we see that the fact that children of divorced parents do not get divorced more often suggests that one does *not* learn from a divorced mother that divorce is permissible. This seems to destroy the notion of a subculture. Since American blacks are at least partially assimilated into American society and show in most respects relatively little cultural difference from whites, it would be inconceivable that all blacks would endorse the view that families should be matrifocal. The violent reaction to the Moynihan report demonstrates how offended black leaders were by the accusation that the black family was unstable. But if the mother-dominated family is not the predominant cultural norm among blacks, it might by a minority norm, perhaps part of a "lower-class subculture." If it is a subculture, however, then it follows that many of those who have unstable marriages must be members of the subculture and transmit the subcultural values to their children; but that is not what happens.

One might wonder at the lower divorce rate reported by men as compared to women. Some of this difference can be accounted for simply by that fact that men are older than their wives. The age of breakup of first marriage would be lower for women then men, and many of the women now in their thirties and forties would be divorced from men who are now over 45 and hence not in our sample. There is probably also some misreporting, by mothers who claim a divorce but have in fact not married, or by men who say they are single when in fact they are divorced. For that matter, given the presence of common-law marriages, whether one wishes to say one is married or not is a matter of judgment. (To the extent that men deny having been married, this supports the contention that divorce is disapproved of.)

If marital instability is not transmitted from parents to children, but is instead endemic to black life in a more general way, then even if we were able to maintain marriages in one generation, the number of unstable marriages in the next generation would not be greatly reduced. In the rest of this chapter, we will try to learn something about the situational factors which cause marriages to break up.

The Role-Segregated Marriage; Additional Measures of Marital Satisfaction

Mirra Komarovsky, in her book *Blue-Collar Marriage,*[20] argues that lower-class marriages are characterized by a segregation of roles; the husband derives important gratification from activities with "the boys," in sports, or at the corner bar, while his wife spends her happiest hours with the wives on the block. The couple communicates little and is unable to relate intimately. While our data contain no measure of intimacy, we do have a listing of leisure activities, and each respondent was asked whether he or she did these things and whether they did them with or without their respective spouses.[21]

In general, about two-thirds of the time devoted to such leisure activities as going out to eat, to the movies, for a ride, or visiting friends, are done by husband and wife together. The pattern of participation is consistent; the respondent who does one activity without his or her spouse is likely to do others alone as well. Seven of the items—going to a movie, eating in a restaurant, taking a drive or walk for pleasure, having a drink in a bar, playing cards, visiting friends, and having a long conversation—were combined into a single scale. The average γ between the seven items is .65; and the association between eating out alone and going to a movie alone is .85. Thus the seven usable items make a good scale measuring the extent of social participation with one's spouse.

The extent of participation with spouse provides us with a second measure of the quality of the marriage. Role segregation is associated with the "best times" indicator used earlier in this chapter. The 16% of our respondents who said "without their spouse" when asked "Since you have been married, do you think you have had your best times when you were doing things with your husband (wife) or when you were with friends but without your husband (wife)?" are of course more likely to score low on the index of participation with spouse; the gammas between the two measures are .59 for women and .30 for men. Our third and most direct measure of marital happiness is the response to the question, "If you had your life to live over again, would you marry or not?" Twenty-seven percent of our married respondents answered "no."

The respondents with role-segregated marriages—i. e., low participation with spouse scores—are less likely to say they would marry again (γ = .24 for men, .21 for women). Finally, men and women who say that their best times have not been with their spouses are much more likely to say they would not marry

---

[20] Mirra Komarovsky (1964), *Blue-Collar Marriage,* Random House, New York.
[21] Item is taken from Norman M. Bradburn.

again; these gammas are .68 and .52. These three measures are used in the analysis that follows to measure marital happiness.

Age at First Marriage

Divorce rates among those who marry very young are extremely high. This is true for whites, and Table 9.11 shows this for blacks. Since women marry earlier than men, we have broken the age distributions differently—early marriages for men are those aged 19 or less, for women it is age 17 or less. Table 9.11 (right-hand column) shows that 25% of the men and 28% of the women married at these early ages. Actually this overstates the situation, since a number of our respondents have not married yet; when they do, they will be older at first marriage, so the final percentage of first marriages which are youthful will be less than the percentages shown.

Table 9.11 indicates that over half of the early marriages have already been broken; as our respondents grow older, a few more will be terminated. In contrast, men and women who married later have much more stable marriages.

Similarly Table 9.12 shows that those youthful marriages which did not break up are not as happy as marriages in which the couple was older. All of the gammas are in the predicted direction.

Social Status and Marriage

If we consider only our measures of marital happiness and postpone for a moment the question of divorce, we see in Table 9.13 that high status persons generally have happier marriages. Table 9.13 lists five indications of socioeconomic status—the respondent's parents' education and his own education, personal and family income. We do not compute personal income of women who are married since they presumably are not the primary breadwinners. Using the three measures of marital happiness, there are 27 measures of association between socioeconomic status and marital happiness. Of the 27, 22 are positive. Most of the associations are small for men, but in the case of women several of them are of moderate size. For example, two of the three associations with education are above .30. The third measure of marital happiness, proportion of leisure activities spent with spouse, does not correlate well with education for women; but this variable is poorly correlated with all the other measures of social status for women as well. This is probably because the decisions regarding leisure activity are made by the husband—if she goes out alone, it is because he has decided not to involve her in his social life.

When we turn to the question of separation and divorce however, the pattern becomes more complicated. First of all, the associations are generally higher

TABLE 9.11
Age at First Marriage and Outcome of First Marriage

Age at first marriage	% divorced or separated	% of all first marriages at this age
MEN		
19 or less	51 (360)	25
20-21	27 (371)	26
22-24	23 (290)	20
25 or older	16 (411)	29
	$\gamma = -.62$	100
WOMEN		
17 or less	58 (519)	28
18-19	40 (579)	31
20-21	37 (344)	18
22 or older	29 (417)	22
	$\gamma = -.40$	99

TABLE 9.12
Age at First Marriage and Marital Success of Those Still Married

	Association with age at first marriage for respondents who are still married (γ)	
	Men	Women
Best times with spouse	.08	.23
Would marry again	.14	.14
Large % of activities with spouse	.10	.07

for men than for women. Two of the associations, with the education of each parent, are distinctly negative for women. There seems to be no obvious explanation for these two negative associations. Better-educated women are less prone to divorce, as the table indicates. Since the education of the parent is a reasonably good predictor of the respondent's own education (see Chapter 5) one would assume that the woman with the well-educated mother would herself have more schooling and hence be less likely to have her marriage break up. One possible explanation is that women from well-educated homes are forced to marry downward in social status because of the shortage of middle-class men. This could place an additional strain on their marriages. The difficulty with this explanation is that the women with well-educated parents are more happily married, even though their divorce rate is high.[22]

[22] Possibly women with well-educated parents have fewer "concealed" divorces because they are involved in fewer common-law marriages.

TABLE 9.13
Socioeconomic Background and Marital Success (γ)

	MEN				WOMEN			
	Best times with spouse	Would marry again	Leisure activities with spouse	Not divorced or separated	Best times with spouse	Would marry again	Leisure activities with spouse	Not divorced or separated
High mother's education	.04	−.02	.08	.21	−.11	.08	.02	−.19
High father's education	.14	.02	.09	.21	.17	.21	.05	−.12
High Respondent's education	.03	.11	.19	.19	.32	.36	.11	.12
High respondent's personal income (men only)	−.05	.06	.13	.24	*a*	*a*	*a*	*a*
High family income	−.05	.04	.14	*a*	−.17	.24	.15	*a*

[a]These associations are not computed since they are either irrelevant or misleading.

The association between social status and divorce for men is actually much stronger than it appears in Table 9.13. The reason for this is that a minority of men from high-status backgrounds marry early, and these marriages tend to break up. If we limit ourselves only to men who marry after the age of 20, the effect of social status becomes sharper. If the marriage is an early one, the men with well-educated fathers are more likely to be divorced rather than less likely ($\gamma = +.31$ between father's education and divorce). While this finding is not statistically reliable due to the small number of cases, it does seem plausible. Many of these men must have married before they completed the college program which their parents had in mind for them. They may have been rebelling against their parents and jumped impulsively into marriage. When we look at late marriages, we see that men with well-educated fathers are much less likely to get a divorce ($\gamma = -.69$).[23] This means we have located one group of men for whom the "black family" problem does not exist—men who grew up in stable middle-class or upper working-class families and avoided the temptation to enter into a teenage marriage (we say "stable families" because relatively few children in broken homes know their father's education). The problem is that only a small number of people fall into this category. Only 9% of the respondents had fathers they knew were high school graduates, and one-fifth of these had teenage marriages. So our low divorce rate group included only 7% of all black men.

The respondent's social background is almost as good a predictor of his marital stability as is his present social status. The man's income and education are moderately good predictors of his marital stability but the education of his parents is equally good. Thus it looks as if social class is important for men, partly because it is easier to make a marriage work if there is enough money, but also because being raised in a middle-class home is apparently good preparation for marriage. In addition the middle-class boy, being less rebellious, should internalize the norms of the society more completely and hence be less likely to see divorce as a legitimate solution to his problems.

Divorces are Initiated by Men

Recall that in Table 9.13, the various background characteristics correlated poorly with measures of marital happiness for men and correlated well for women. Conversely, background characteristics predicted marital breakup better for men. There is a theory of divorce, consisting of three assumptions, which would fit this pattern. Let us use education as an example, bearing in mind that the assumptions are general enough that any other background trait or person-

---

[23] The 5% confidence interval for this gamma is .43 to .80.

ality characteristic could be used instead. The three assumptions are: (1) the well-educated respondent is happier in his marriage not because he or she receives better treatment from the spouse but because the better-educated person is more tolerant, less demanding, and in general happier. The respondent's education affects his own sense of marital satisfaction—his preception of the marriage relationship—more than it affects his spouse's perception of the marriage.

(2) Women are more completely involved in marriage than men; consequently a wife will be more preoccupied with the evaluation of her marriage, less willing or able to ignore dissatisfactions and unmet needs. Men, on the other hand, have more opportunities to derive satisfaction from outside the home—through work, friends, or even extramarital sexual behavior. For these reasons the correlation between personal characteristics and marital satisfaction will be higher for women than for men. The poorly educated man has opportunities outside of the marriage to meet his needs. The poorly educated woman, confined primarily to the role of wife and mother, is less able to compensate for her sense of dissatisfaction in the marriage.

(3) Men have the economic freedom to leave their families; women generally do not have the freedom to leave their husbands. Thus the poorly educated woman will remain in an unhappy marriage; the poorly educated man will leave.

Taken together, these three assumptions will lead us to predict that education is an important correlate of marital happiness for the wife but not the husband; it is an important predictor of the man's marital stability, but not the woman's.

We cannot prove this complete theory, but it is true that almost all the data we have seen so far are consistent with it. Consider Table 9.10 which related family stability to divorce rate. The table showed three differences which support this interpretation: (1) coming from a home broken by either death or divorce causes marital unhappiness; this effect is stronger for women than men; (2) coming from a home broken by either death or divorce causes higher divorce rates; this effect is stronger for men than women; and (3) coming from a home broken by divorce apparently inhibits divorce in a way that coming from a home broken by death does not; this effect is also stronger for men than for women.

Let us now look at other kinds of variables, to see if this pattern continues. In Table 9.14 we repeat four of the variables we have already looked at: the respondent's education, the education of each parent, and age at first marriage. We have added four of the five basic psychological variables (expression of anger, internal control, self-esteem, and happiness) and also region of birth. We have omitted fight and arrest rates, because we think these characteristics are not so obviously causes of divorce and may well be effects of divorce. In the first two columns of Table 9.14, we present the average association between the variable and two measures of marital happiness—the "best times with your spouse"

TABLE 9.14

Associations of Various Factors with Marital Happiness and Differences Between the Sexes in Marital Breakup

	Average γ two measures of marital happiness		Difference of absolute values	Separation or divorce (γ)		Difference of absolute values
			Men-women			Men-women
	Men (1)	Women (2)	(3)	Men (4)	Women (5)	(6)
High education	.07	.34	−.27	.19	.12	.07
High father's education	.08	.19	−.11	.21	−.12	.09
High mother's education	.01	−.02	−.01	.21	−.19	.02
Born in North	.08	.04	.04	−.03	−.05	−.02
Older at first marriage	.11	.18	−.07	.62	.40	.22
Recall anger	−.09	−.25	−.16	.07	.01	.06
High internal control	.24	.34	−.10	.22	.17	.05
High self-esteem	.03	.04	−.01	−.03	−.01	.02
High happiness	.10	.42	−.32	.33	.32	.01

question and the answer to "would you marry again?"—first for men, then in the second column, for women.[24]

For example, in Table 9.13 we saw that, for men, the gamma between mother's education and "best times with spouse" was + .04, and the gamma between mother's education and "would marry again" was −.02. Averaging these two γ's, we get the quite insignificant value of +.01. Repeating this arithmetic for women, we get the equally uninteresting value of −.02. In column 3 we enter the difference, subtracting the absolute values of the entry in column 2 from the absolute value of the entry in column 1. (The absolute value is simply the γ without its negative sign, if it has one.) Thus if the mother's education has more impact (either favorably or unfavorably) on the wife's happiness than on the husband's happiness, the difference in column 3 will be negative. In this example, we see that we get −.01 in column 3—too small to pay attention to. But if we look down column 3, we see that the signs are consistently negative; with only one exception, each factor affects the woman's marital happiness more than the man's. The difference between the γ's is over .10 in five instances.

[24] We cannot use the third measure of marital success—participation with spouse—because we assume that wives have less control over this aspect of the marriage. And the fact that this measure correlates well with background for men is consistent with the theory: those aspects of the marriage which men control (going out and divorce) should correlate better for men.

In column 4 and 5 we present the associations of these variables with ever having been divorced, and take the difference of the absolute values in column 6. Column 6 shows a consistent set of small positive differences, indicating that the man's divorce rate is more influenced by these factors than the woman's. Of the ten associations, only one is negative; this is region of birth, the same item which produces the deviant sign in column 3.

All of this seems to support our contention that men initiate the majority of black divorces. But there is an interesting exception to this. College-educated women also initiate divorces. In Table 9.15, we look again at two of our best correlates of divorce—internal control and coming from a broken home—and divide men and women into three educational levels: those who did not finish high school, high school graduates with no college, and those with one or more years of college. The pattern is similar for both variables; the correlations are positive and large for men at all levels, close to zero for women in the two lowest educational levels, and similar to men for college-educated women. It seems likely that divorce is less painful for college women than for others since they have the least to lose economically from divorce. Women from broken homes are less satisfied in marriage, and unhappy college women, having the economic resources, will elect divorce where more poorly educated women cannot. Exactly the same argument applies to the internal control associations.

The fact that college-educated women apparently initiate divorce does not mean that they have higher divorce rates. As we have already seen, the higher the level of education the lower the divorce rate for both sexes. Education must have a very strong stabilizing effect on marriage for blacks since the divorce rate is so low among well-educated couples, despite the fact that both of them are free to initiate a divorce.

The Psychological Causes of Marital Instability

Taking a last look at Table 9.14, we see a number of factors that are important predictors of divorce. The most important seem to be happiness, internal

TABLE 9.15

Associations of Internal Control and Happiness with Marital Stability for Men and Women at Different Educational Levels (γ)

	Men			Women		
	Less than high school	High school graduate	College	Less than high school	High school graduate	College
Internal control	.32	.24	.22	.01	.06	.30
Broken home	.13	.18	.26	.01	.02	.33
N	(853)	(383)	(261)	(1068)	(597)	(349)

control, education, parents' education, and coming from a stable home. Of these, happiness cannot be considered a cause of divorce. Our evidence is that men and women who have remarried are actually slightly happier than the average man and woman still in their first marriage. But if happiness were a completely unchanging personality trait and the cause of divorce, then those men whose unhappiness had driven them to divorce would still be unhappy after remarriage. It looks as if the association between happiness and divorce should be interpreted as merely meaning that people who live alone are less happy than those who are married.

We have not included the fight and arrest variables because we cannot determine with certainty the causal relationship between fight and arrest and divorce.

Taken together, this set of correlates portrays the likely candidate for divorce as a lower-class man who is aggressive, demanding, often depressed, who lacks some of the interpersonal skills for marriage and has too weak a commitment to society's norms to resist the attraction of divorce. This means that the factors which cause blacks to divorce are probably not too different from the factors which produce breakups in white marriages.

In addition to the effects of poverty and social class, however, there must be an additional (racial) factor; for poor blacks have higher divorce rates than equally poor whites. To the extent that discrimination and prejudice are causes of low internal control, or antisocial aggression, then discrimination and prejudice are the indirect causes of marital instability. Part of what is happening may be that "being treated like an inferior" increases various frustrations and needs which the husband brings into his marriage and which then help to destroy it. (We will consider the merit of this line of argument in the next chapter.)

Summary

The effects of coming from a home broken by divorce, separation, or death are not uniformly bad and depend on such factors as region (North or South) and sex. Coming from a broken home *does* depress income for men and educational attainment for women. Poverty does not account for the effects of the broken home, nor does the hypothesis that males from "female-dominated" homes underachieve. Personality variables are strongly associated with the broken home: men from broken homes tend to be more violently aggressive and less happy, and to have a low sense of internal control. These factors, in turn, are associated with the respondent's own marital stability.

The data indicate that the man with low education, low internal control (and perhaps a need to express aggression) initiates divorce; apparently women, unless they are college graduates, do not initiate very many marital breakups.

10

Discrimination, Personality, and Achievement

If this study were done on a racially homogeneous population, then we would expect to find that "social class" or "socioeconomic status" is the major factor correlated with most of the characteristics we observe. We would expect, in general, that middle- and upper-class people possess more of those attributes responsible for social or economic success and that the poor are poor by virtue of having less of these attributes. Furthermore, the advantages of the middle and upper classes would not be solely economic, but would also be attitudinal and psychological. In short, the "advantaged" would not only be financially secure, better fed and housed, healthier and more educated than the "disadvantaged," they would also be more industrious, self-confident, emotionally secure, and happy.

These latter characteristics may be construed as consequences of higher social "status." The middle-class person may of course begin with superior genetic endowment, but beyond that he is socially and emotionally conditioned for success. From childhood he is taught that he has the power to manipulate his environment to get what he wants and that the world is generally a friendly place. In contrast, the lower-class person views the world from the bottom, as the used rather than as a user. Lacking the material means for success, he does not expect to succeed and therefore finds himself at a psychological disadvantage as well.

This aspect of the difference between the middle and lower class also applies to whites and blacks with one crucial difference: the lower-class person can resolve to "try harder" and achieve middle-class status, the black cannot hope to become white. Whether the black person views the world from the bottom or top depends less on how hard he works or how strongly he believes he can succeed than on the relative *caste* positions of whites and blacks. If we then complicate the situation by making the exact relationship between the castes ambiguous and in a state of rapid social change, we make the individual's position even more uncertain. In this case, each person is free to make his own assessment of the caste situation and to define his own position within it.

The thesis of this book is that the caste system in this society is a variable much like that of social class in a racially homogeneous society. The black who perceives the caste sanctions bearing upon him as being relatively strong and inflexible will see himself hopelessly at the mercy of the "system" that dominates him. On the other hand, the black who perceives these caste sanctions as being weak or surmountable will see himself much the same way a middle-class white sees himself—as someone capable of succeeding for himself. Of major concern in this book is the black who falls into the former category. Our reasoning is that the black individual who is treated as an inferior and who believes himself to be helpless will develop certain personality traits which prove to be dysfunctional for his personal success. This is generally the argument advanced by Kardiner and Ovesey and by Martin Luther King, Jr., although without the kind of data available to us. Our task is to find a mode of analysis that will enable us to test these relationships.

Measures of "Being Treated Like an Inferior"

The proper research design would require a comparison between two groups that are genetically similar but subject to different environments. We cannot compare blacks to whites without first demonstrating that the two groups are identical in all ways except their caste position. Ideally, we have simply to match American blacks with another group who have lived in a very similar industrialized western nation with complete equality for several generations. (While we are being facetious, we might propose selecting randomly a large number of blacks, transplanting them to some other nation, and studying them again several generations later.)

Ignoring these textbook approaches to the testing of the theory, what can we do? The best alternative available to us is to separate blacks into those who have experienced the least sense of "being treated like an inferior" from those who have experienced the most mistreatment. Admittedly this is a poor alternative. In fact, all blacks are in large measure treated the same; they are segregated and discriminated against in similar ways. But the best we can do is try to separate blacks in this way, and test four variables which should measure differential treatment.

1. *Number of years lived in the South.* Only 23% of our respondents were born in the North, and 14% came North before they were 10 years old. Clearly,

the South is more repressive than the North, and therefore native Northerners should think and act differently due to growing up in a less restrictive society.

2. *Skin color.* Since blacks are identified by the color of their skin and since miscegenation has been so common in our society, then it makes sense that the darker the skin, the "more black" the subject will feel. Many different writers have argued that lighter skin is a "status-symbol" to blacks, or at least was a generation ago. In 1940 Charles S. Johnson interviewed 2000 black adolescents, and found that 48% of the girls described "the boy you like most" as brown or yellow; but only 19% said "the ugliest boy you know" was light. Conversely, 42% of the "ugliest boys" and only 4% of the boys most liked were black. Girls were slightly more color-conscious than boys. The most respected and admired boys and girls were light-skinned, and this was especially true if the criterion used was sexual attractiveness.[1]

In theoretical terms, it seems likely that the power and privilege of whites makes them subjects of admiration, and therefore the physical characteristics of whites are preferred.[2] This in turn means that the light-skinned black will like himself because he looks white, and use bleaches and hair-straighteners to look whiter. For the same reason it is likely that having white ancestry was admirable in the recent past (and perhaps still is). Skin color is measured in this survey by merely asking the black interviewers to rate the black respondent on a six-point scale, from "very light" to "very dark."

3. *Childhood Segregation.* Being "treated like an inferior" means, in part, to be segregated. If we are correct that attitudes are developed from childhood experience, then those blacks who experienced an integrated childhood should have different attitudes than those whose early experiences were segregated. We have several items in the questionnaire which were used to form a cumulative scale: whether the respondent attended an integrated elementary school, whether he lived in an integrated neighborhood, whether he played with white children and had white friends, and whether the parents entertained whites in their homes.

Despite myths to the contrary, southern black children are much more segregated than Northerners; northern-born respondents were more likely to attend integrated schools, live in integrated neighborhoods, play with whites, and see their parents relating socially to whites. Table 10.1 shows the percentage of the respondents who had various integrated childhood experience.

Older blacks experienced less segregation than younger ones. Primarily this is because the ghettos were smaller, and therefore, there was a greater chance of whites

---

[1]Charles S. Johnson (1941), *Growing Up in the Black Belt,* Chapter 10, Schocken, New York.

[2]This is sometimes called identification with the aggressor, and is the basis for Jewish concentration camp victims sometimes copying the dress of German guards, as reported by Bruno Bettelheim (1961) in *The Informed Heart,* Free Press, Glencoe; see also Stanley M. Elkin (1968), *Slavery,* Univ. of Chicago Press, Chicago.

TABLE 10.1
Contact with Whites by Region of Birth and Age

Past contact with whites	Born in the North (%)		Moved to the North before age 10 (%)		Moved to the North after age 10 (%)	
	21-30 years	30-45 years	21-30 years	30-45 years	21-30 years	30-45 years
Attending integrated elementary School[a]	48	62	43	41	3	3
Lived in integrated neighborhood[b]	77	86	74	84	62	68
Played with white children "often"	73	85	70	82	46	53
Parents had white friends visit them "frequently"[c]	18	30	6	20	15	18
N	(646)	(672)	(256)	(306)	(660)	(1573)

[a]A respondent is considered to have attended an integrated elementary school if he attended a school which he said was "half white" for 5 or more years.

[b]Integrated means more than half white.

[c]This question was asked only of children who lived in integrated neighborhoods, hence it may understate the actual amount of visiting by excluding parents in segregated neighborhoods who had white visitors.

living nearby.[3] The 30- to 45-year-old respondents celebrated their sixth birthdays sometime between 1927 and 1942—before the heavy northern migration created by the war. However those in their twenties did not turn 6 until after the war began. (This no doubt explains why relatively few older blacks are in the "moved North before age 10" category; this would have meant moving North during the Depression.)

4. *Socialization.* The final variable is the respondent's report of what his parents taught him about racial etiquette. Most of what a child learns about being black he learns from his parents rather than through personal experience.

Children begin making racial distinctions before they enter school. Kenneth and Mamie Clark developed an experimental technique in which children were

[3]One easy way to understand the relationship between interracial contact and size of the ghetto is to note that those who live on the perimeter of the ghetto have more opportunities for neighborhood contact with whites. If we assume that the ghetto is roughly circular, and a narrow band around the boundary is integrated, then the area of the ghetto is approximately proportional to the square of the perimeter. If the ghetto quadruples in size, the perimeter increases only twofold, and a smaller proportion of the total area is on the integrated edge.

given brown and white dolls and were asked to select the one they wanted to play with and the one most like them. Three-year-old children were unclear on their own color, but the 5-year-olds tended to give correct responses. Unfortunately, the 5-year-olds also showed preference for the white doll, sometimes with straightforward comments rejecting the brown doll—"he is ugly" or "she is bad."[4] With older children, the white doll was more likely to be rejected. This suggests that the racial attitudes of young children are critical, and hence we drafted a series of questions on the way in which the respondent's parents introduced the child to the race issue.

Logically, the critical task of the parents' socialization would be in helping the child in a tense racial situation. Did the child learn at that time that his parents could help him in some way? Or did they only make things worse by demonstrating to the alert child that they were as powerless as he to cope with the situation?

First, we asked the respondent to recall an incident in his childhood when a white adult caused him some trouble because of his race, and then asked, "What did your parents say about that happening?" Only one-half of the respondents were able to recall an incident. For those respondents who could not recall an unpleasant childhood incident we asked, "Did your parents ever tell you what you should do when whites were nasty or rude to you?" The answers were then "coded" into various categories in three different ways: first we considered what sort of advice the child was given about how to act, either in this particular incident or other situations that might occur; second we coded the response according to whether the parents' philosophy or point of view came through in the response; and third we simply coded the response by asking ourselves whether the parents' statement gave support to the respondent's status or whether it gave the impression that whites were somehow of higher status, better, or more powerful than blacks.

We have subdivided each set of codes into those that we considered "active," those we considered "passive," and those which are neither or are uncodable by this standard.[5] An active response is one which enhances the respondent, defends him, gives him some reassurance. Most commonly, an active response would reassure the respondent that it is all right to be black, that he should treat all adults respectfully, but not be uncomfortable when he is around whites. One respondent (whose father was a respected southern educator) recalls it this way: "They tried to explain that this is the way it was at the time, and someday it would improve . . . . [When an incident occurred] I should exercise good judgment as

---

[4] This work is summarized in Kenneth and Mamie P. Clark (1955), *Prejudice and Your Child,* Beacon Press, Boston.

[5] Coding was done by Charlene Jones Allison from a coding scheme developed by David Klassen.

I [react] to the event . . . I was taught to be polite to *all* my elders." Another respondent, born in Philadelphia, said his parents told him not to differentiate between whites and blacks. "They just told me to be courteous to all adults. They did tell me not to say 'Sir' or 'Ma'am' to anybody." Both responses are reasonably supportive; the first exemplifies what we have called "the appropriate response"—to defend oneself, but only if it is warranted. There are other responses in which the offending white is "put down." Another southern high school principal's son remembers being told, "Some people just don't want to accept others as human beings, but they are ignorant and have to be taught. . . ." But these answers are exceptional, even in the North. More commonly, the response seems to be a brief admonition to forget about it, which would seem to be easier said than done. Many respondents indicate that color was not something one could talk about openly. As one said: "They didn't tell me anything. It was a known fact how you are supposed to act. You saw how your parents acted and you did the same." And of course in the South, there is the frequent response, "Always say 'yessir'." When incidents occur, the respondent is told to stay away from whites. The most common single response to an incident was simply to forget it or ignore it. These responses we have coded as "passive" since they imply that there is nothing to be done, no way to protect one's self-esteem.

The question "How were you told to act around whites?" has two main kinds of replies: (1) to be deferential to whites, or (2) to be deferential to all adults, black or white. We have coded the first response as passive, the second as active. Then the four different codes were combined into a single variable called "parental socialization." When this is done, approximately four-fifths of the respondents can be scored from "very passive" to "very active," and only 20% of the cases are unusable.

Perhaps one should be surprised at the almost total absence of hostile responses. However, long years, first of slavery, then of heavy-handed oppression, have effectively stymied open rebellion. The Panthers did not exist when this survey was taken, and today that group has only a tiny following. A larger number (frequently estimated at 15% of the adult population in riot areas)[6] took part in a riot, but this is a spontaneous, once-in-a-lifetime excursion into rebellion. The everyday language of our respondents shows little hostility. Rather we see in these responses a mixture of rather humanitarian idealism and outright passivity.

These, then, are the four factors we will examine: region of birth, skin color, childhood contact with whites, and parental teaching regarding race. The simplest demonstration of our thesis would be to show that those blacks who have been

[6]Robert M. Fogelson and Robert B. Hill (1968), "Who Riots? A Study of Rioters" in *Supplemental Studies for the National Advisory Commission on Civil Disorders,* p. 231, U. S. Government Printing Office, Washington, D.C.

most oppressed according to these four indicators have the greatest difficulty. However we have already seen enough data to know that this rather simple test will fail. Southerners are most oppressed by these measures; in addition to being southern by definition, they are darker in color, more segregated, and more passively taught. But we know that divorce and crime are problems for the northern-born, and that southern migrants have higher incomes than Northerners once educational differences are statistically removed. Either our hypothesis is wrong—freedom is no help—or else there is some ugly complicating factor which makes life worse for Northerners and cancels out the gains which increased freedom brings.

In the rest of this chapter we will analyze the effects of region of birth, skin color, parental socialization, and segregation. In order to place the results in perspective we will also look at mother's education, our best measure of parental social status. Region of birth is highly correlated with the other measures of racial treatment, so it will be necessary to carry out separate analysis for northern- and southern-born respondents. Otherwise we are fortunate in that within each region the various measures are not highly intercorrelated, as Table 10.2 shows. Those who attended integrated schools score high on the "childhood contact with whites" scale, but they are not higher status, lighter colored, or more actively socialized. There is some tendency for Southerners with well-educated mothers to be more actively socialized and lighter, but in general we can analyze the variables independently of each other without any statistical problems.

Region of Birth

The North, at least by comparison to the South, is a land of freedom and opportunity. This sense of freedom is communicated at an early age to the child —the northern-born child is much more "actively" socialized.

But the net effect of being raised in the North is unfortunate, as Table 10.3 shows. Northern-born men and women have considerably higher self-esteem. The γ for men of .41 indicates that self-esteem is pushed up more by northern birth than any other factor. Ability to express anger is also high among Northerners. They are less anti-white and show more internal control. On the other hand, Northerners are less happy and more likely to fight or be arrested. Thus when we remember that northern-born respondents are better educated and have better-educated parents, the γ of .17 between region and internal control looks unimpressive. Certainly caste sanctions are less severe in the North. Discrimination is not backed by the full weight of law, as it was in the South when our respondents were growing up, and a northern black can speak out without fear of being lynched. Apparently the theory that inhibition is a response to the repression of blacks is correct, for when this inhibition is reduced in the North, self-

TABLE 10.2
Association Between Background Factors (γ)

	Integrated high school	Integrated elementary school	High childhood contact	Light skin	Actively socialized	High mother's education	Stable home
MEN							
Born in the North[a]			.61	.13	.45	.58	.07
Integrated high school	X	.79	.51	-.13	-.13	.05	-.11
Integrated elementary school		X	.95	-.03	-.11	.03	-.05
High childhood contact			X	.08	.14	.15	-.06
Light skin			-.02	X	.06	.14	.17
Actively socialized			.03	.13	X	-.06	.04
High mother's education			-.05	.25	.24	X	-.15
Stable home			-.09	.01	.10	-.05	X
WOMEN							
Born in the North[a]			.57	.31	.49	.45	-.04
Integrated high school	X	.80	.65	.16	.35	.07	.25
Integrated elementary school		X	.96	.13	.04	.22	.12
High childhood contact			X	.17	.02	.12	.14
Light skin			-.02	X	.03	.13	.10
Actively socialized			.36	.02	X	-.06	.11
High mother's education			.13	.13	.11	X	.19
Stable home			-.09	-.01	-.02	.07	X

[a]Northern-born above diagonal, southern below.

TABLE 10.3
Region of Birth and Personality (γ)

	Association with northern birth	
	Men	Women
Recall anger	.27	.24
Self-esteem	.41	.25
No fight/arrest	−.04	−.09
Internal control	.17	.06
Happiness	−.02	−.10
Low anti-white	.07	.20

esteem and expression of anger increase. Using our measure of self-esteem, northern blacks do not have lower self-esteem than whites. At the same time, being born in the North does not increase the security factor very much. There are two explanations for this. First, the fact that an individual can express feelings more freely means he can admit that white prejudice exists (instead of avoiding the issue by denying the existence of prejudice), and he may even blame whites for his failures which actually are not the result of discrimination. Secondly, while the boundaries of proper behavior are well-defined in the South, the North presents a much more ambiguous picture—a situation characterized by normlessness. This ambiguity can produce a sense of *anomie*—of not knowing where one stands. It can also mean that an instance of discrimination which would be taken for granted in the South comes as a surprise in the North and causes more unhappiness. Thus, while white prejudice is not more severe in the North, the northern respondent may be more aware of it and more depressed by his awareness.

Socialization

Part of the importance of region is the behavior of parents: southern black parents communicate to the child by their words and actions what the world is like. The difference between the North and South are reflected both in what parents say and what effect it has on the child. For example 58% of the southern respondents said that their parents cautioned them to be deferential (say "yessir") around white grown-ups, compared to only 25% of the northern-born respondents. What one's parents said should be important in forming a conception of what it means to be black. If the respondent was given a "passive" socialization, we can expect him to grow up more afraid of whites, perhaps more inhibited, and and with greater problems controlling his aggression. This is precisely what happens.

In Table 10.4, we see that actively socialized women and southern men are more secure, but not northern-born men. Active socialization is most strongly associated with happiness; the gammas of .17, .32 and .30 suggest that parental

TABLE 10.4
Patential Socialization and Personality

| | Association with Active Socialization (γ) | | | |
| | Women | | Men | |
	Born in the South	Born in the North	Born in the South	Born in the North
Recall anger	.07	.03	−.01	.09
Self-esteem	.14	.40	−.03	.08
No fight/arrest	.06	.24	−.10	.01
Internal control	.07	.14	.20	.04
Happiness	.17	.32	.30	−.19
Low anti-white	.18	.07	.21	−.05

socialization is one of the more important causes of happiness. In Chapter 7 we argued that the stable, long-term component of happiness was probably more than anything else simply the absence of internal anxiety. We also saw that blacks were much less happy than whites, and it makes sense that being introduced to the race issue by parents who are supportive should lessen anxiety. A principal assumption of our analysis is that subconscious fear of whites is a major problem for blacks, and these data are certainly consistent with that point of view.

A similar pattern occurs with internal control. For Southerners, being told that it is all right to be black, that whites have no right to be overbearing, that one can expect life to get better, should lead to a greater sense of mastery of the environment. But when we look at northern-born men, we see that socialization has no relationship to control. Similarly, being actively socialized reduces anti-white attitudes for Southerners and for women but not for northern men.

In the northern ghettos, active socialization does not mean merely teaching children to hold their heads up in the face of persistent white oppression. It may mean the open expression of feelings about whites which are too dangerous to express in the South, and it may mean teaching what we would call prejudice if we were talking about white children. Being allowed to express hostile feelings about whites will not build one's confidence that whites will be fair, or that the environment can be trusted and controlled. Such behavior can, in effect, reinforce the original feelings that whites are unfair and that they control one's life. Thus freedom from the regularly enforced caste regulations of the South is not enough. Nor is being able to freely express one's feelings without fear of reprisal enough. The sense of internal control also requires the knowledge that one really can strive to achieve in the white man's world. Perhaps it also requires the knowledge about what whites are really like, which can only come from face-to-face contact with white peers. Freedom without hope is no help.

This difference between North and South is also reflected in the fact that active socialization was characteristic of well-educated parents in the South, but in the North it was associated with parents who were poorly educated.

If active socialization is detrimental because it releases too much aggression, then it no longer seems inconsistent that it should be a problem for northern men but not for northern-born women. Women are less aggressive than men; reducing their inhibition will be less likely to result in pushing them over the line into overexpression of aggression.

Skin Color in the North and South

Perhaps the breakdown of inhibition in the North has begun to lead to the disappearance of skin color as a relevant social variable. We expected to find that light-skinned blacks were more emotionally secure, and this does seem to be the case except again for the northern-born men. (Table 10.5).

Traditionally, light-skinned men and women held higher prestige. However a study by Udry, Bauman, and Chase[7] of a Washington, D.C., sample showed that over the past decade there has been a steady decline in the preference for light-skinned men as marriage partners (but no change in the case of women).

This is consistent with our tabulation; light-skinned men born in the South and light-skinned women act as if they have higher status; they have more internal control, are happier, are less anti-white, and have less trouble with aggression. They are also slightly less assertive.

For northern men, the pattern does not appear; northern-born light-skinned men are not more secure.

Childhood Contact with Whites

The most obvious form of mistreatment which blacks receive is segregation. How does the black child react when he sees that none of the other children in

TABLE 10.5
Skin Color and Personality: Association (γ) with Light Skin Color

	Women		Men	
	Born in the South	Born in the North	Born in the South	Born in the North
Recall anger	−.17	.10	−.04	−.15
Self-esteem	−.08	−.04	−.09	.01
No fight/arrest	.23	−.12	.23	−.04
Internal control	.22	.13	.24	.07
Happiness	.12	.16	.15	−.09
Low anti-white	.05	.15	.08	−.03

[7] J. Richard Udry, Karl E. Bauman, and Charles Chase (Jan. 1971), "Skin Color, Status, and Mate Selection," *Am. J. Soc.,* 7, 722-733.

his neighborhood or school are white? Does this affect his personality development? The answer (Table 10.6) is that it has a decisive impact on Northerners, especially men—the exact opposite of what we have seen up to now.

For northern men 11 of the 12 associations between security and high school integration, elementary school integration, and childhood contact (a summary scale including contact in school and neighborhood and childhood friendships) are positive, and 7 of the 12 are above + .15. Most impressive are the associations with happiness: 29% of northern-born men who attended segregated schools are unhappy, compared to 14% of alumni of integrated elementary schools. Or to put it another way, the average alumni of an integrated elementary school is considerably happier than is the graduate of a segregated school with an income of over $8500! No other variable—parents' education, coming from a stable home, owning one's own home, etc.—is as powerful a predictor of happiness for northern-born men.

Apparently contact with whites, in the North and on an equal-status basis, is necessary to prevent a black from growing up with serious anxieties about race. Happiness is closely linked to internal control, as we saw in Chapter 8, and this is consistent, since contact with whites also increases internal control for northern-born men.

We often assume that the northern integrated school is integrated in name only, and that attending such a school is a frustrating, even traumatizing experience. Similarly, we assume that an integrated neighborhood is a racial battleground. Hence one might assume that attending an integrated school might increase anxiety, since one would experience prejudice and mistreatment. However the data indicate that in the North more unfortunate incidents occur to children in segregated neighborhoods. It is the white children from outside one's own neighborhood who are the danger. (For example, students who attended northern segregated schools are more likely to remember an incident of being mistreated than are those who attended integrated schools.)

E. Franklin Frazier's *Youth at the Crossroads*[8] describes the life of black teenagers in the border states 30 years ago, and gives many examples of racial tension. However, even in this rather unpleasant environment, he does not give any examples of children in integrated play groups being victims of prejudice. For example, one child remarked:

> I never thought I was white, but when I was little I never played with anything but white children and I thought I was it. When I was a little girl the white children around me never did call me "nigger" but I remember when I was seven or eight the white kids in the next square did. You didn't have to do anything to them. All you had to do was walk down the street.

[8] E. Franklin Frazier (1967), *Youth at the Crossroads*, Schocken, New York.

Table 10.6
Childhood Contact with Whites and Personality (γ)

| | Men | | | | Women | | | |
| | Southern-born | Northern-born | | | Southern-born | Northern-born | | |
	Contact	Contact	Elementary school integrated	High school integrated	Contact	Contact	Elementary school integrated	High school integrated
Recall anger	.13	.07	−.15	−.08	.27	.18	.06	.30
Self-esteem	.13	−.01	−.13	−.04	.24	.07	.01	.09
No fight/arrest	−.12	−.06	.02	.30	−.19	.02	−.05	−.05
Internal control	.04	.11	.18	.16	−.06	.06	.09	.10
Happiness	−.11	.32	.43	.39	−.18	.16	.16	.17
Low anti-white	.05	.12	.16	.08	.11	.11	.06	.18

If a white child plays with a black, the black is valuable to him as a playmate. No matter what the culture says, the black child has the power to simply play elsewhere, and he therefore can resist being mistreated. Consider this story related to Frazier:

> We used to play movie actresses. We got along all right until one day we played an "Our Gang" picture. They wanted me and Jane to be the little colored children. I didn't think anything about it but Jane said no. We would not play it unless one of them would be a colored child. They said we could play it better than they could because we looked more like colored children than they did. Still Jane wouldn't play. Finally one of the white girls said she would be a little colored girl.

Clearly this black child has learned a lesson: she can successfully make a demand on whites.

The explanation, then, is that contact with whites at an early age leads the child to realize that whites are not as much the enemy as he feared, that they do not all hate him, that he belongs to some degree in their world. Conversely, northern segregation teaches him not so much that he is bad, as that whites are bad, and he grows up feeling anger and frustration.

The data on women fit into this model if we assume that black women are more afraid of whites than black men are. Then integration, in both the North and South, will reduce their fear and permit an increased expression of aggression. This is precisely what the data show in Table 10.7. If we assume that southern women are more inhibited than southern men, who are in turn more restricted than northern-born men, we see in Table 10.7 that the more inhibited a group is, the greater increase in assertiveness results from integration, and the greater loss in sense of security. Integration makes southern women most militant and northern-born men more moderate. In the extreme case of northern men, integration in elementary school has an inhibiting effect on assertiveness. Perhaps this is because the assertiveness shown by segregated Northerners is an overcompensation—a strutting—which becomes less necessary when security is increased.

TABLE 10.7
Relationship Between Region and Sex and the Effect of Integration

Category	Level of inhibition	Effect of childhood contact in increasing assertiveness (gammas with anger and self-esteem)	Effect of integration in increasing security (gammas with internal control and happiness)
Southern-born women	Very high	Positive (.27, .24)	Negative (−.06, −.18)
Southern-born men	High	Weak positive (.13, .13)	Mixed (.04, −.11)
Northern-born women	Low	Weak positive (.18, .07)	Weak positive (.06, 16)
Northern-born men	Very low	Mixed (.07, −.01)	Positive (.11, .32)

One way to think of this pattern is that American society in its race relations is somewhere in the middle on a continuum from slavery to equality. Slaves are not openly unhappy; and people with full citizenship have less to be unhappy about. But in between, a shift toward equality should produce both a cause to be happy and an opportunity to express one's anger. If we imagine how a group would move along this continuum, it seems plausible that the first step, the precondition for further advancement, would be to overcome fear, and replace it by anger. This raises that possibility that desegregation of schools in the South may lead to increased militancy. Blacks in integrated schools might become more assertive and less secure, with all the attendant problems for both blacks and whites which this would cause. But on the other hand it may be that school desegregation will enable the South to bypass the long period of violence and low achievement which the North is now going through. It seems impossible to predict what will happen. Whatever may happen in the future in the South, at this time the most unhealthy situation seems to be that produced by northern ghettos, where discrimination, in the form of segregation, is obvious, and where blacks are free from the day-to-day supervision of whites so that they may freely express their anger.

Thus the hypothesis that discrimination is a major factor in producing black-white personality differences, seems correct, although not in the simple way in which it was initially assumed. In only one way do the four measures of "being treated as an inferior" have the same effect. All four (southern-birth, dark skin, passive socialization, and segregation) cause the subject to be more anti-white. In all we have seen 18 γ's between a racial factor and anti-white attitudes, and 16 of the 18 are positive (the exceptions are dark skin color and passive socialization for northern men). The highest, .21, is between passive socialization and anti-white attitudes for southern-born men.

Integration and Parental Social Status as Causes of Achievement

If integration produces such large effects in increasing the security of northern men, the next question is whether integration thus has a long-term impact on achievement. To answer this, let us contrast the impact of school integration with the effect of coming from a high-status family background and the effect of the respondent's own education.

Coming from a family where the mother is a high school graduate tends to raise both the respondent's assertiveness (i.e., self-esteem and willingness to recall anger) and his security as Table 10.8 shows. This is to be expected, since both assertiveness and security are increased by higher levels of education, and respondents with well-educated parents are themselves better educated.

Again, the effects are weak for northern-born men. For the other three groups the effects are sizable. For women, the effects of mother's education on internal

TABLE 10.8
Mother's Education and Personality (γ)

	Women		Men	
	Southern-born	Northern-born	Southern-born	Northern-born
Recall anger	.22	.07	.12	.19
Self-esteem	.27	.28	.35	.02
No fight/arrest	.12	−.04	.10	.16
Internal control	.22	.22	.31	.12
Happiness	.13	.30	.05	.14
Anti-white	.05	−.03	.09	.29

control are as strong or stronger than any of the racial variables. Northern-born women with well-educated mothers are also happy. High mother's education pushes women's self-esteem up as much as being born in the North does. For southern men, the effect of mother's education on internal control is large, but the other associations, and the associations for northern-born men, are small.

It is interesting that the one group for whom integration is most important is also the group for whom background social status is least important.

Perhaps it is reasonable that northern men, more than any other group, should be sensitive to racial factors and insensitive to social status. In the more stable order of the South, where competition against whites is outlawed, a stable status structure among blacks has developed, and there is some advantage to a high social position. In the North this sort of status means much less.

Social Status, School Integration, and Achievement

Table 10.9 represents the effects of school integration, mother's education, and respondent's education on the eight achievement variables used in Chapter 8.

By including respondent's education as both a column and row variable, we can see to what extent the background variables act through increasing education to produce an effect.

Table 10.1 showed that attending an integrated school was not strongly associated with parental social status. The γ between mother's education and elementary school integration is .22 for women born in the North, .03 for northern-born men. This seems surprising; we would expect middle-class parents to have more chance to send their children to integrated schools. But in fact, even middle-class respondents do not have very much control over whether their child's school is integrated. Integration is most often the result of accidents of city size, level of black and white demand for housing, and local school attendance rules. For example, the best predictor of integration is size of city; only 20% of the respondents who attended northern schools and are presently living in metropolitan areas of under 2,000,000 population attended black-majority schools, compared to 37% of those living in larger metropolitan areas. In Minneapolis, even today, the black population is too small to fill a segregated high school. There are only 5255[9] black students in all 12 grades in Minneapolis, hence only four of the city's 98 schools—three elementary and one junior high—are predominantly black. In 1940, when many of our respondents were in elementary school, the black population of Minneapolis was a quarter of what it is today. It would have been extremely difficult to create a predominantly black school under those circumstances. Even in cities with somewhat larger black populations, historical accidents sometimes work to prevent segregation. For example, Boston had no high school which was over 70% black in 1968. The main reason for this seems to be that the underfinanced Boston School Committee lacked funds to construct a high school in the Roxbury ghetto. Thus any of our respondents who grew up in Boston must have attended integrated high schools. On the other hand, the respondent growing up in Chicago would have had difficulty obtaining an integrated schooling.

There is a certain amount of parental choice, of course. One of the most interesting ways in which this appears is in the high number of women respondents who attended integrated schools. Apparently parents look to "good neighborhoods" in order to protect their daughters. The boy who had no older sisters had only a 45% chance of attending integrated schools; the son with older sisters had a 58% chance. Similarly, the daughter who was preceded by older sisters is more likely to have gone to an integrated school. This seems strong evidence that choice of school was affected by the sex of the children.

On balance, there is a limited amount of free choice in attending integrated schools, especially in high school, but it does not seem likely that it is large enough to vitiate the apparent effects of integration we are about to see. If voluntary integration were a large factor, children with well-educated parents would be much more likely to attend integrated schools, and they are not.

In general we see that school integration has a positive impact on achievement. Twenty-five of the 26 γ's are positive, and 15 are over .15. The only negative one is the association between coming from an integrated elementary school and being divorced for men, which is inexplicable.

The associations in Table 10.9 have been controlled for age, since older respondents had more opportunity to go to school with whites.[10] Integrated schooling, especially at the high school level, seems to be associated with high levels of home ownership and high scores on the finance scale. The association of high school

---

[9] *Directory of Public Secondary and Elementary Schools in Selected Districts, 1968,* Dept. of Health, Education, and Welfare, Office of Civil Rights.

[10] The statistic in Table 10.9 is a partial γ, a weighted average of the γ's computed separately for older and younger respondents. See Appendix 5.

TABLE 10.9

Mother's Education, Respondents Education, and School Integration as Causes of Achievement, Controlling on Respondent's Age

	Association (γ) with achievement							
	High education	Not divorced	Tried for integrated house	Own home	High finance score	Low job change	High income	Know of job
Northern-born:								
Elementary school integrated								
Men	.04	-.18	.17	.31	.03	.06	.11	.06
Women	.07	.09	.30	.03	.24			
High school integrated								
Men	.38	.17	.26	.26	.46	.30	.21	.01
Women	.14	.07	.18	.18	.30			
High mother's education								
Men	.19	.21	.20	-.08	-.09	-.09	-.08	.08
Women	.24	.16	.26	.13	.41			
High respondent's education								
Men		.15	.32	.60	.59	.21	.32	.43
Women		.33	.22	.29	.49			
Southern-born:								
High mother's education								
Men	.45	.16	.12	.19	.38	.20	.15	.21
Women	.49	.10	.18	.16	.37			
High respondent's education								
Men	.13	.13	.05	.12	.49	.23	.29	.21
Women	.13	.13	.18	.39	.51			

integration with home ownership was analyzed in more detail, and no control variables could be found which reduced this association.[11]

The forms of achievement which seem to result from integration for Northerners are the ones associated with the "security factor"—home ownership, financial responsibility, and job stability, in particular.

This is completely consistent with our earlier finding that alumni of integrated schools are much less unhappy. Apparently something occurs in the integrated school which reduces anxiety and makes it easier for a black man to direct his energies toward certain kinds of achievement. The fact that men from integrated schools do not know more about job opportunities is also consistent with the low assertiveness of northern integrated men. Finally, the fact that the effects are generally stronger for men than for women agrees with the pattern for the effect of integration on personality.

Taken together, this seems convincing evidence for our general thesis. The black personality traits which conservatives like to single out as the explanation for black poverty are indeed causes of poverty; but the ultimate cause of black maladjustment, anxiety, and unhappiness is the oppressiveness of the caste sanctions which blacks must endure. And for the present generation of blacks, northern birth coupled with integrated schooling seems to compensate in part for this.

In order to see the magnitude of the effects of integration, they can be compared with the effects of the respondent's educational attainment and the effects of his parental background—here measured by mother's education. Respondents' education is in nearly every case a stronger correlate of achievement than is integration. A graduate of a segregated high school is more likely to achieve than a dropout from an integrated high school. But this is a misleading way to put it, since attending an integrated high school sharply decreases one's chances of dropping out. Mother's education is a better contrast to integration, and in general we find that school integration has more impact on achievement than does parental social status. Mother's education is more strongly associated with high school graduation than is elementary school integration; and it is definitely associated with marital stability, trying to find integrated housing, and job knowledge. A well-educated mother apparently has more assertive children.

The effect of mother's education on other forms of achievement is fairly strong for women and for southern-born men, but for northern-born men it is surprisingly weak. The associations with home-ownership, financial security, job tenure, and income are weakly negative. Again, the groups most strongly affected by integration are least affected by social status. Despite all the great emphasis on "cultural deprivation," childhood social status is not a major factor for northern black men. In the South, where a stable social hierarchy exists among blacks, mother's education is more important.

[11] Analysis carried out as a class project by Hopkins graduate students.

Parents' socioeconomic status is also more important for northern women than for northern men. This is consistent with the finding in Chapter 8 which showed several indicators of social status highly correlated with happiness for women. For example, home ownership is correlated .63 with happiness for northern women, but only .35 for northern men; scoring high on the finance scale is associated .44 with happiness for northern women and only .19 for northern men; never being divorced and happiness are correlated .40 for northern women, .25 for northern men. The fact that these associations are stronger may reflect the fact that women are more inhibited than men—more southern in their personalities. But it may also reflect the fact that to the black women, whose life may be centered heavily on her family, the home may be an important symbol. Women tend to say they are happy if they are satisfied with their marriage: the γ between happiness and "would marry again" is .26; between happiness and "best times with spouse," the γ is .76. For men, there is *no* relation between happiness and the responses to these two questions. If the woman has a job or a career, it is probably in the low-pressure world of feminine occupations, where there is less chance for advancement and less racial discrimination. Thus for both the housewife and working woman, relations with whites are probably less critical. But if black-white relations are less critical, more conventional factors such as social status can be more meaningful. For a woman the differences between being single and being happily married, between being a clerk and being a teacher, or between renting an apartment and owning a home, are real enough to provide a basis for believing that one need not be angry at the world.

But for the black man, none of these things may be very relevant. His main role in life is to work, and by working, to achieve a position of respect, and to provide security for his family. To work is to "keep the wolf from the door." But his position in the world of work is defined and bounded by the operation of white society. He can have no more than he is permitted. Or, as William C. Berry, Director of the Chicago Urban League, said:

> I think that the most important single factor that can be known about an individual in these United States is not his talent, his character, or even his money, but his color—even when one gets high up. Let me give you an example. All of you, I expect, have heard the name Dr. Percy Julian, world-renowned chemist, who has worked on cortisone, rubber-based paint, and so on. He is a Negro, a Ph.D., leads a very productive life, is rich, and has a pretty wife. He bought a house outside the ghetto, and the white people tried to burn the damn thing down. Now, what more in terms of American requirements for success, can one have besides a Ph.D. and be rich and an inventor and having a pretty wife? He has nothing else coming. And they still want to burn the man's house down just because he is brown.[12]

[12] Symposium, *Daedelus*, (Winter 1966), p. 311.

Thus the black man needs more than anything else to know what his place in society is. Yet there is no such thing as *the* place of a black man in this society. There are many places, and not every house in the suburb is burned. But every man simplifies his world, glosses over many of the complexities, generalizes so as not to be overwhelmed by detail. And the key to the way in which a person generalizes about the world is his perception of what it means to be black. Whether the world is fair or not is less important than whether one *thinks* it is fair.

We have thus succeeded in providing a partial answer to the questions we posed in Chapter 2. Why is the North such a bad place to grow up in? Why do southern migrants have higher incomes and experience less difficulty than native Northerners? The answer seems to be that the North provides freedom but that freedom in a segregated ghetto means merely freedom to despair over one's predicament, freedom to blame whites, freedom to act out the "black rage" that the Southerner has learned to hide so well.

Since this survey was done in 1966, before the wave of riots and before black power became highly fashionable, we did not plan to investigate the new militancy. However, without particularly intending to, we have learned something about its genesis. The data have documented something which many writers have said: that northern blacks in the big ghettos are as anti-white and as pessimistic as southern blacks, but with more willingness to express these feelings. Militancy then serves as a useful mechanism for release of this hostility, and it is an ideology which is compatible with the high self-esteem of northern black men.

The question remains, is it functional? Is black militancy an ideology which can encourage individual achievement? Our data won't answer this question, but there are some associations in the data which are discouraging. First of all, young Northerners are more anti-white than older Northerners, as we would expect. If the black power argument were valid, these younger men would combine this anti-white feeling with high self-esteem, but also with high internal control. When we look at the data however, we find that young Northerners have lower control scores, and anti-white scale scores are as highly correlated with low internal control for young men as for old. This suggests that the new militancy has not found a way of separating expression of hostility toward whites from despair over one's chances of success.

The new militancy (by which we mean an endorsement of confrontation, hostility toward whites, and black separatism) is not therapy, a prescription for change; posing as therapy, it is merely a conservative defense of the attitudinal status quo which northern blacks were practicing in 1966. It may be viewed as the logical transformation which could be expected to occur when the civil rights movement left the inhibited South and came to the assertive northern ghettos. It is usually said that black militancy is built on the failure of the civil rights movement. But in one sense, it is equally reasonable to say it is built on its success. For one cannot attack whites unless one is fairly sure one can do so with impunity. The most extreme groups, such as the Panthers, do run the risk of imprisonment and death, but the average nonviolent militant does not; he may even have had the opportunity to test his new militancy on his white friends to make sure it was acceptable.

We began this analysis assuming that the black with strong anti-white feelings was being realistic; those who were not anti-white were being defensive and denying the truth. The question, "How cruel are whites?" is not a scientific question, of course. But our instinctive belief that whites are very cruel made it easy to understand that Southerners were being defensive and internalizing aggression in a dysfunctional manner, and made it much harder for us to understand that northern blacks were also being defensive in externalizing their anger, and perhaps of the two bad choices, internalization was the wiser.

Summary

Black-white personality differences have their roots in the differences in the way society treats the two races. Blacks raised in the South, particularly those with dark skin whose parents socialized them to be accepting of the race situation, appear as adults as inhibited, unhappy, with a low sense of internal control, and are more anti-white. Northerners, who do not have to grow up under the oppressive sanctions of southern society, are much less inhibited. However, if they remain isolated in the ghetto, they may remain unhappy, with a low sense of security. Actual peer contact with whites in childhood, primarily through school integration, does reduce anxiety and increase the respondents' sense of security. This is reflected in the higher achievement of alumni of integrated schools. Contact with whites in the South has mixed effects. It tends to reduce black inhibitions, but it also lowers scores on the security variables.

The North and the South are very different. In the South our respondents grew up in, discrimination was openly practiced, and blacks had no political freedom—not even the freedom to vote. In the North, discrimination is officially frowned upon, but more important, blacks do, theoretically, have the basic political freedoms—freedom to vote, to organize, freedom of speech and of the press. But for most of our respondents, this political freedom means merely the rights to recognize that he is confined to a ghetto.

In the South, interracial contact without political equality makes life worse: the southern black would be better off if he never saw a white person. In the North, there is no social equality; so that political equality alone, without integration, is also destructive. Freedom without hope is no help.

11

School Integration: An Evaluation

At the end of Chapter 10, we examined the effects of school desegregation, which is the only aspect of integration presently a real issue of public policy. It is also one of the few places in civilian life where there is enough experience with integration to permit a definite evaluation.[1]

The Supreme Court's famous 1954 opinion included the following paragraph:

> Segregation of white and colored children in public schools has a detrimental effect upon the colored children. The impact is greater when it has the sanction of the law; for the policy of separating the races is usually interpreted as denoting the inferiority of the Negro group. A sense of inferiority affects the motivation of a child to learn. Segregation with the sanction of law, therefore, has a tendency to retard the educational and mental development of Negro children and to deprive them of some of the benefits they would receive in a racially integrated school system. . . . Whatever may have been the extent of psychological knowledge at the time of Plessy vs. Ferguson, this finding is amply supported by modern authority.[2]

The decision then footnoted seven sources in the social science literature, ranging from Myrdal to a sample survey of the opinions held by social scientists on the validity of the theory that segregation depresses achievement. These data were clearly inadequate as evidence and served largely as "expert opinion" rather than statistical proof.

[1] There are several analyses of racial integration in the military.
[2] The text of the Court's decision and of the social scientists' brief for the N. A. A. C. P. are printed in Kenneth Clark (1955), *Prejudice and Your Child,* Beacon, Boston.

154

It was 12 years before the federal government was willing to finance the necessary research. First the Office of Education funded *Equality of Educational Opportunity* (the Coleman report).[3] A year later, the Civil Rights Commission financed a study consisting primarily of a reanalysis of the Coleman data by James McPartland.[4] The commission also funded the collection of the data for this book, some tabulations of which appeared along with the McPartland analysis in *Racial Isolation in the Public Schools.*[5]

The Coleman data showed convincingly that integration resulted in improvements in the achievement test scores of black students. Most of this effect could be attributed to the fact that the white students in the school were from higher status backgrounds, but some of it is also a purely racial effect, i.e., black students in integrated schools do better than those in equally middle-class all-black schools. Coleman's analysis[6] suggests that the overall gain in verbal ability, for blacks in integrated schools is approximately one-fourth of a standard deviation, or slightly less than one grade level. McPartland, in his analysis of the same data, argues that the true difference is one-half of a standard deviation. Since the difference between the average white and the average black is about one deviation, this means that one-half the white-black difference could be removed by integration.

Our study represents the first time that an attempt was made to evaluate the effects of school integration on people after they had completed their schooling. This means that our evaluation can look at more than achievement test performance; it can look at the number of years of school completed, adult occupation and income, and adult personality and attitudes toward whites.

Integration and Educational Achievement[7]

Each respondent was asked how many years he had attended elementary school with white students and approximately how many white students there were in his school. The respondent is considered to have attended an integrated school if he said that (1) he attended elementary school with whites for 5 or more

---

[3] James S. Coleman, Ernest Q. Campbell, Carol J. Hobson, James McPartland, Alexander M. Mood, Frederic D. Weinfeld, and Robert L. York (1966), *Equality of Educational Opportunity,* U. S. Office of Education, Washington, D.C.

[4] James McPartland (1968), "The Segregated Student in Desegregated Schools," The Center for the Social Organization of Schools, Baltimore. See also McPartland (1969), "The Relative Influence of School and of Classroom Desegregation on the Academic Achievement of Ninth Grade Negro Students," (Summer) *J. Soc. Issues, 25,* 93-102.

[5] U. S. Civil Rights Commission, *Racial Isolation in the Public School* (1967), U. S. Government Printing Office, Washington, D.C.

[6] Coleman, *et al., op. cit.,* p. 332.

[7] The interested reader is referred to Robert L. Crain (1971), "School Integration and the Academic Achievement of Negroes," (January) *Soc. Educ.,* where this topic is treated in more detail.

years, (2) the school was at least half white, and (3) the whites did not move out of the school while he was attending it. The definition of integrated high school experience is the same, except that there is no time restriction.

Table 11.1 gives the association between integration and number of years of school completed. In general, respondents from integrated schools have more education. For example, the first column in the top panel gives the educational attainment for northern-born men whose education was entirely in integrated schools (this means either the elementary school and high school were both integrated, or else the elementary school was integrated but the respondents did not go to high school). The clearest comparison is shown in the third column (men whose elementary and high school education, if any, were segregated). Nearly half of the segregated respondents did not finish high school (36% + 12% = 48%). The figure is only 36% (28% + 8%) for alumni of integrated schools. Integration seems to reduce the dropout rate by one-fourth.

In the same fashion, male migrants from the South who attend northern segregated schools are more likely to drop out than those attending integrated schools. (Compare in the middle section, the 20% + 57% = 77% dropout rate for segregated schools to the 12% + 33% = 45% dropout rate for integrated schools). Integration seems to cut the dropout rate for southern migrants nearly in half. The lower part of the table shows the same pattern for women, although the effect is much smaller.

Turning to college attendance, we again see a consistent tendency for integration to mean higher rates of college attendance. For example, 32% of northern-born men from integrated schools went to college, while only 24% of northern-born men from segregated schools did so. Among southern migrants, the differences in college attendance is enormous: 30% versus 3%. (Hopefully, some of this difference is sampling error.) Again, the differences for women are in the same direction, but they are very small. Finally, we have too few college graduates in the sample to establish a clear pattern, but in three cases out of four, graduates of integrated schools are more likely to finish college.

The mixed category refers to respondents who went to a segregated elementary school but an integrated high school, or sometimes the opposite. Since a respondent has to go to high school in order to have a mixed school experience, there are no elementary school dropouts in the "mixed" category. The "mixed" respondents seem more like the integrated group than the segregated one; their dropout rates are low, and their rates of college attendance are high. If we analyze the effects of elementary and high school integration separately (Chapter 10), we find that the apparent positive effect of integration on high school graduation rates is all due to integration of the high school, and elementary school integration makes little difference. Elementary school integration does make a difference in rates of college attendance, however.

TABLE 11.1
Educational Attainment by Integration of Elementary and High School, Birthplace, Age Moved North, and Sex

Educational attainment	Born in the North			Born in the South, moved north before age 10			Born in the South, moved north after age 10	
	Both schools integrated (%)	Mixed (%)	Both schools segregated (%)	Both schools integrated (%)	Mixed (%)	Both schools segregated (%)	Mixed (%)	Both schools segregated (%)
MEN								
Less than 9 years	8	a	12	12	a	20	a	37
9–11 years	28	23	36	33	29	57	48	32
12 years	32	45	28	24	51	20	24	19
Some college	27	22	20	18	17	0	24	8
College graduate	5	10	4	12	3	3	3	4
N	100	100	100	99	100	100	99	100
	(273)	(158)	(159)	(66)	(70)	(60)	(58)	(852)
WOMEN								
Less than 9 years	6	a	16	11	a	30	a	26
9–11 years	37	48	35	50	8	36	44	34
12 years	32	31	25	25	39	22	47	26
Some college	13	11	20	10	41	9	8	8
College graduate	12	10	5	3	21	3	8	5
N	100	100	101	99	101	100	99	99
	(310)	(157)	(153)	(107)	(96)	(69)	(106)	(1097)

[a]Respondent must attend high school to have a "mixed" education.

For comparison purposes, we have included (in the far righthand column of Table 11.1) data on respondents who received most, and in most cases, all of their education in the South. A very large fraction of the southern-educated women have more schooling than southern-educated men. The same sex difference appears in northern segregated schools, but in northern integrated schools, men usually have more education than women. The southern school and the northern segregated school seem to reward girls more than boys.

In summary, Table 11.1 points out three things:

1. Alumni of integrated schools are more likely to finish elementary and high school and to attend and finish college.

2. Respondents who attended integrated high schools and segregated elementary schools fare as well in terms of finishing high school as those whose schooling was entirely integrated.

3. The effects of integration are stronger for men than women. Put another way, the southern school and the northern segregated school both show a bias in favor of women, while northern integrated schools show a bias toward men.

The effect shown in Table 11.1 is large. If 10 children were sent to integrated schools instead of segregated ones, one of the ten would finish high school instead of dropping out, and another would go to college instead of stopping his schooling with a high school diploma. Any educational innovation that can claim that result is pretty remarkable. But we are getting ahead of our story, since we have not yet established that integration is the real cause of the effect shown here. Let us consider one other educational output, a verbal achievement test score, and then control on the effect of family background to show that it is not the case that integrated schools simply recruit better students to begin with.

Achievement Test Scores

Table 11.2 indicates that students who attended integrated high schools and elementary schools scored higher on the verbal achievement test than do those who attended segregated schools. In three of the four cases there is a consistent trend, with respondents from all integrated schools scoring highest, those from mixed schools second, and those from all segregated schools lowest. Looking at the top line of the table, however, we see that the effect for males is small. For male migrants from the South, it is slightly negative. Since the more schooling respondents have, the higher they score on the test, the difference for males could be due entirely to the fact that males in integrated high schools are less likely to drop out.

There is a large difference between the test scores of integrated and segregated females, and the difference is too large to be due only to differences in educational attainment. Apparently girls in integrated schools not only stay in school slightly longer, but more important, learn more while they are there.

TABLE 11.2
Verbal Test Score by Integration of Elementary and High School, Birthplace, Age Moved North, and Sex[a]

Mean test score	Born in the North			Born in the South, moved north before age 10			Born in the South, moved north after age 10	
	Both schools integrated	Mixed	Both schools segregated	Both schools integrated	Mixed	Both schools segregated	Mixed	Both schools segregated
MEN	28.6	27.8	27.5	28.5	29.4	28.8	26.0	24.1
N	(280)	(153)	(176)	(64)	(70)	(68)	(68)	(786)
WOMEN	30.8	30.1	26.6	26.3	25.8	25.0	24.8	24.1
N	(343)	(157)	(165)	(105)	(92)	(73)	(106)	(1065)

[a]σ of test scores = 8.5; effect of integration significant at .05 level (one-tailed).

The largest difference in the table is between respondents educated in the North and those educated in the South. Again, this difference is larger than we would expect solely on the basis of Northerners having more years of schooling.

Family Background and Integration

Before we can consider the reasons why school integration has this strong effect on educational achievement, we should disprove the obvious counterhypothesis that blacks in integrated schools are from superior backgrounds. In Chapter 10 we argued that the association between mother's education and integration was too low to explain the apparent effects of integration. Let us now take time to demonstrate this.

Table 11.3 summarizes part of the relationships in Tables 11.1 and 11.2 by presenting the differences between the percentages who are high in educational attainment and verbal achievement when the high school was integrated and when it was segregated and the same difference in verbal test score due to elementary integration. Beneath these three differences, we have presented the percentage difference which remains attributable to integration when father's education, mother's education, number of siblings, and stability of family or origin are each controlled in turn. These "net partial percentage differences" are simply the average effect of integration in the table after the background variables have been introduced as controls. (See Appendix 5) As the table indicates, the net differences are only slightly smaller than the gross differences, indicating that the original association cannot be explained by these other variables.

Surprisingly, black students in integrated schools do not come from higher-status families than those in segregated schools. Considering only northern-born

TABLE 11.3
Association of School Integration with Educational Attainment and Verbal Test Scores with Background Variables and Sex Controlled

Additional variables included	High school integration on educational attainment (%)[a]	High school integration on test score	Elementary school integration on test score
None	+10.6	+1.4	+2.1
Sex only	+10.8	+1.3	+2.0
Sex, father's education	+10.9	+1.3	+1.5
Sex, mother's education	+ 9.6	+1.2	+1.4
Sex, number of siblings	+ 9.8	+1.4	+2.1
Sex, stability of parental family	+10.1	+1.2	+1.4

[a]This column is the increase in percent graduating from high school. Other columns are increase in test mean.

respondents, 42% of alumni of integrated elementary schools had mothers who completed high school, compared to 38% of alumni of segregated schools. Women alumni of integrated schools are more likely to come from unbroken homes, 58% compared to 44%. This tendency for integrated respondents to come from stable homes is too small to have much effect on the associations between integration and achievement, as Table 11.3 indicated. This is a report of conditions when these respondents were of public school age, but apparently this situation has not changed since then. Coleman found that parents' education was correlated $-.006$ with integration at the 9th grade; the parallel correlation with presence of the father in the home is $-.06$.[8]

Integration and Occupational Achievement[9]

School integration also affects adult occupational behavior. Alumni of integrated schools are more likely to move into occupations traditionally closed to blacks; they also earn slightly more money, even after education is controlled.

For this part of the analysis we will look only at the effects of high school integration on black men.

Table 11.4 records the percentage of blacks for each of the eight major urban occupational groups in the 1960 census and the percentage of alumni of integrated high schools and of segregated high schools in each of these occupations. Black men tend to be concentrated in the lower blue-collar occupations—operatives, service workers, laborers—and in the lowest of the white-collar occupations —clerical work. Conversely, black men tend not to be professionals, managers, salesworkers, or craftsmen. Black alumni of integrated schools show a definite tendency to go into these traditionally white occupations; approximately one-third of the male black alumni of integrated high schools are in crafts, sales, and the professions, while only one-fifth of the blacks who attended segregated schools are in these three groups. Male black managers, owners, and proprietors tend to come from segregated schools, but contact with whites is not necessary to enter these occupations, since almost all black managers are in businesses serving a largely black clientele. (If the data were available, we would hypothesize that black businessmen serving white clientele would be more likely to have had integrated schooling.)

One reason black clerks are slightly more likely to be from segregated schools is that black clerical positions are available in the largest metropolitan areas where there also is the largest number of segregated schools. When city size is

[8]Coleman, *op. cit.,* Appendix Vol. 2, Appendix, p. 239.

[9]The interested reader is referred to Robert L. Crain (1970), "School Integration and the Occupational Achievement of Negroes" (January), *Amer. J. Soc.,* **78,** 593-606, for a slightly more detailed discussion.

TABLE 11.4

Occupations of Men from Segregated and Integrated Northern High Schools by Sex,
and Percentage of Blacks in each Occupational Group[a]

Occupational group	Blacks in group (%)	Integrated (%)	Segregated (%)	Difference
Professional	5.9	11	8	+3
Managers, owners, proprietors	4.2	3	6	−3
Sales	4.7	3	0	+3
Craftsmen	7.5	19	13	+6
		36	27	
Clerical	11.3	11	13	−2
Operatives	14.3	31	40	−9
Service	29.1	15	10	+5
Labor	27.4	6	10	−4
		99	100	
N		(498)	(227)	

[a]Alumni of southern high schools excluded from this table.

introduced as a control variable, the apparent predominance of men from segregated schools in clerical work becomes smaller.

The eight major occupational groups are broad categories, and we can continue this investigation by looking at differences in the detailed occupational
classifications within each major group. In Table 11.5 we see that men from
integrated schools are more likely to hold those occupations within each major
occupational group whose work force is less than 3% black. We shall also refer
to these detailed occupational groups which are less than 3% black as nontraditional. For example, 36% of black professionals from integrated schools are in
the nontraditional professions, compared with 33% of black professionals from
northern segregated schools. (Of course, in Table 11.4 we saw that blacks from
segregated schools are less likely to be in the professions at all.) This is the
smallest difference in the table; the other differences are sharp even for such low-
level employees as operatives and service workers. Managers, operators, proprietors, and laborers have been dropped from the table because there is no classification of managers, etc., which is over 3% black and no classification of laborers
which is under 3% black. The clerical and sales categories have been combined.

These two tables yield convincing evidence that black alumni of integrated
schools are in "integrated" jobs.

Not surprisingly, blacks in nontraditional jobs are better educated and have
higher incomes. But educational differences will not explain income differences.
If two respondents have the same education, the one in the nontraditional job
will have the higher income.

TABLE 11.5

Percentage of Respondents from each Major Occupational Group in (Detailed)
Occupations which are less than 3% Black, by Integration
and Region of High School

| Major occupational group | High School | | |
	North, integrated (%)	North, segregated (%)	South, segregated (%)
Professional	36 (56)	33 (18)	31 (35)
Clerical, sales	19 (67)	13 (30)	30 (49)
Craftsmen	56 (93)	41 (29)	15 (113)
Operatives	4 (155)	0 (91)	0 (179)
Service	8 (76)	0 (22)	0 (51)

Table 11.6 shows that blacks who attend integrated high schools have higher incomes.

Various factors need to be controlled before we can arrive at a true estimate of the real dollar return resulting from an integrated education. The two most important factors to control are age and stability of home. Young men earn less than older men; but since the extent of school segregation has been increasing in the North, young men are also more likely to be segregated. This means that the "real" effect of education is less than $344. On the other hand, men from broken homes are more likely to go to integrated schools, which tends to artificially depress the effect of integration on income. When both of these factors are controlled, the effect of integration increases slightly to $390. Part of this income gain is due to the higher education of men from integrated schools; when education is added as a control, the effect of integration becomes $210 per year. This estimate may be too high, and we do not have enough northern-educated men to make this estimate reliable. The effect of integration (net of education) may be as low as $100.

Clearly, a larger sample is needed to make this estimate; but until one appears (which, we suspect, will not be soon) we must assume that integration has a net effect on income, independent of other variables including the higher educational attainment which also results from integration, of about $100 per year or more—not a small difference over the 40- to 50-year working life of an adult male.

There are three main ways in which integration affects black achievement. One we have already discussed at length: the effects of contact with whites on the personality. A second way is the effect of the differences in norms, values, and behavior of classmates in integrated schools. We will call this the "contextual" effect of integration. The third mechanism is the effect on achievement of developing friendships and social contacts with whites; let us call this the "interactional" effect.

TABLE 11.6
*Median Income of Alumni of Segregated and Integrated
High Schools (males only)*

High School		Difference
North, integrated	North, segregated	
$5454 (493)	$5110 (247)	$344

Contextual Effects

The white students in integrated schools are less likely to be delinquent, less likely to be slow learners, less likely to drop out, and more likely to go to college, than is the average black student in a segregated school. The black student who goes to an integrated school will conform to the values and norms of his school, and therefore be less delinquent, less likely to drop out, more likely to go to college. He will also learn more simply because the school work is set at a faster pace.

This raises the question: At what racial mix do the "white" norms begin to predominate? One answer is given in Figure 11.1 which plots the percent of black students finishing high school against the number of whites in the school (as estimated by the respondent). The graph shows a sharp gain in percent completing school as the school becomes almost all white; and for girls, there seems to be no integration effect at all if the school is less than half white.

In our data, the only evidence which relates to the contextual effects hypothesis is the impressive association, $\gamma = .42$, between high school integration and response to the question, "Did most of the students in your high school drop out, or did most graduate?" Students in integrated schools are much more likely to recall their classmates' as staying in school. And the association between perception of classmates' dropout rate and whether the respondent himself finished is a very large +.59. This last association is of little value since the student who graduated is probably basing his judgment of the school's dropout rate on his personal circle of friends, rather than the whole school.

The effect of the high school context probably explains why men and women from integrated schools do not marry early. Table 11.7 indicates that both men and women from segregated northern high schools are very likely to have married early, and that early marriage is increasing, since the rates are higher for the younger respondents. For both sexes and age groups the rate of early marriage is much lower among alumni of integrated high schools. We think this simply re-

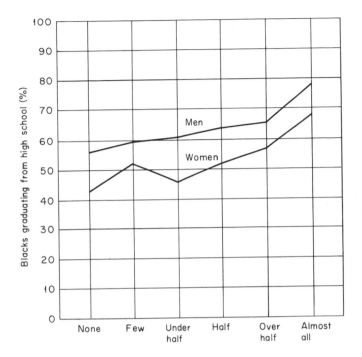

Fig. 11.1. High school graduation, by number of whites in high school and sex.

flects the difference in norms; in integrated schools early marriage is frowned upon by students.[10]

Another example of the contextual effect is the association presented in Chapter 9 between high school integration and never fighting or being arrested, $\gamma = +.30$. This is a large association; but since there is no relationship between fighting or being arrested and elementary school integration, the effect is probably not due entirely to psychological effects of integration. More likely it is simply that boys in integrated schools have less opportunity to be involved in delinquent gangs.

This together with our finding that elementary school integration is unrelated to finishing high school, gives support to the context theory. The low dropout rate from integrated schools is probably primarily the result of school climate, and a black student in a black school with a low dropout rate would probably

---

[10]It should follow from this that graduates of integrated high schools, by avoiding early marriages, should have a lower divorce rate. In fact, divorce rates are not lower for this group, and we do not know why.

TABLE 11.7
Age at First Marriage by Integration of High School

	Northern, integrated	Northern, segregated	Southern, segregated
Respondents under age 30:			
% of ever-married men who married at age 19 or earlier	27 (120)	51 (113)	22 (125)
% of ever-married women who married at age 18 or earlier	24 (245)	32 (130)	22 (195)
Respondents Aged 30–45:			
% of ever-married men who married at age 19 or earlier	13 (256)	30 (99)	11 (280)
% of ever-married women who married at age 18 or earlier	21 (328)	33 (125)	27 (444)

be almost as likely to graduate as one in an integrated school. (Of course, there are very few segregated black schools with low dropout rates simply because there are not enough middle-class blacks to create student bodies with favorable school climates, so this point is of only academic interest.)

The Interactional Effects of Integration

The data in this survey provide a new bit of information about the context theory however. We can see that an integrated high school serves as a foundation for continued interaction with whites throughout adult life. The respondent who went to an integrated high school (or integrated elementary school) is less anti-white, and is more likely to want to live in an integrated neighborhood. As we mentioned earlier, it is not always possible to translate this desire into actually purchasing a house. But graduates of integrated schools are more likely to live in integrated neighborhoods now. To make the clearest test of the hypothesis, we have used elementary school integration, rather than high school. But since students in integrated schools are likely to live in cities with more integrated neighborhoods (which is why they were able to go to integrated schools), this finding was tested separately in each of the 25 cities in our sample. When we do this and restrict ourselves only to northern-born persons, we find that, of a possible 19 comparisons, in 13 cities, those who attended integrated elementary schools are now more likely to be living in an integrated neighborhood. In only six cities is the correlation negative. Furthermore, among the eight largest metropolitan areas, where the number of interviews obtained is enough to provide reliable statistics, we find that the correlation holds as expected in all eight cases.

If we look at southern migrants who were educated in the North, we find the same pattern; of 21 correlations, 14 run in the expected direction, and in the eight largest cities, the correlation is in the expected direction in six cases.

It seems unlikely that this could be attributed to stability of residence; that is, we do not think that very many respondents grew up in an integrated neighborhood, attended an integrated elementary school and continued living in that neighborhood into adulthood. Indeed, there are few neighborhoods in these northern cities which would have remained stably integrated for that many years.

A "present contact with whites" scale was constructed from questions about social contact with whites and neighborhood integration. (The single item in the scale weighted most heavily is "How often do you visit with white friends either in your house or in their homes—frequently, occassionally, rarely, or not at all?"). Alumni of integrated high schools are significantly more likely to score high on this scale. For example, among men, 56% of those who attended integrated high schools scored high, compared to only 32% of male alumni of segregated northern schools and 36% of those who attended high school in the South. The same pattern holds for women (the three percentages are, respectively, 44, 28, and 30%).

Interaction with Whites and Employment

Patterns of social relations should make a difference in job hunting. Respondents were asked: "How did you find your present job (or last job, if not now working)?" When we compare our respondents to the national sample of whites, we find that both groups use informal contacts quite heavily. Only 22% of whites, and 31% of blacks use formal means—newspapers, public and private employment agencies, or their schools. Although only a little more than one-third of the respondents say that family or friends referred them to their present job, another one-quarter of the national sample and one-sixth of the black sample mentioned "visiting plants" as the way in which they found employment.[11] This presumes that the respondent had some idea of what plants to visit, and in a large city, this requires more than a casual knowledge of the labor market. The largest differences between the blacks and the national sample are in this category. It may well be that blacks anticipate discrimination and hence are less willing to make the grand tour of possible employers. Blacks use formal means of obtaining job referrals, such as unions, newspaper advertisements, and public or private employment services more than whites do.

Negroes who attend integrated schools are more likely to associate with whites in later life and have a double advantage in that their contacts with whites also bring them into contact with persons who are better educated. Respondents were asked whether they could go for advice to a relative or friend who was a

---

[11]"These results agree generally with the findings reported in H. L. Sheppard and A. H. Belitsky (1967), *The Job Hunt,* Johns Hopkins Univ. Press, Baltimore.

college graduate. (They were not asked whether the relative or friend was black or white.) Respondents who attended integrated schools are not more likely to have relatives who are college graduates, but they are considerably more likely to have college-educated friends, as shown in Table 11.8. In this table there is essentially no difference among northern-educated respondents who had themselves attended college; in all cases, they were likely to have college-educated friends. But when we turn to respondents who did not attend college, we find that those who attended integrated high schools have very distinct advantages, while alumni of segregated northern high schools are no more likely to have a college-educated friend than are migrants who attended southern high schools. Since alumni of integrated high schools have more white contacts, it seems safe to assume that many of these college-graduate friends are white.

Having friends who are white college graduates should pay off in more information about job openings, and Table 11.9 shows that those respondents who do have college-graduate contacts are considerably more likely to be able to name an employer who would hire them. Notice that the differences are greater for respondents who themselves have some high school or are high school graduates. This is consistent with the possibility that college-graduate contacts and other persons that these respondents could use for referrals would be more familiar with occupations requiring at least minimal educational qualifications.

From these two tables, it appears that the student who does not go to college does benefit from constructing a "referral network" based either on high school acquaintances, or on white contacts he has developed since then.

Table 11.10 closes the argument by showing that graduates of integrated schools who are over 30 years old are much more likely to know of another job. (We have no idea why the age factor should be so important in this table.)

Only one-quarter of our sample stated that they obtained their present job through friends, but this did not mean that the other three-quarters did not

TABLE 11.8

Percentage of Respondents Who Say They Could Seek Advice from a Friend Who is a College Graduate by Integration of High School, Educational Attainment, and Sex

	Northern, integrated (%)	Northern, segregated (%)	Southern, segregated (%)
MALES			
No college	62 (354)	44 (212)	45 (427)
Some college or college graduate	82 (164)	85 (48)	75 (120)
FEMALES			
No college	47 (536)	33 (257)	35 (636)
Some college or college graduate	65 (130)	78 (56)	69 (147)

TABLE 11.9

Knowledge of Another Job Opportunity by Sex, Education, and Contact with a College Graduate (%)

	% Knowing of Another Job, by Education[a]			
Contact with college graduate	Eighth grade	Some high school	High school graduate	Attended college
MALES				
With college graduate contact	32 (217)	37 (414)	41 (365)	52 (332)
Without college graduate contact	29 (184)	17 (185)	21 (122)	(30)
Difference	+ 3	+20	+20	
FEMALES				
With college graduate contact	22 (148)	29 (459)	39 (437)	53 (297)
Without college graduate contact	20 (177)	19 (372)	24 (212)	(58)
Difference	+ 2	+10	+15	

[a]Net effect of college graduate contact, among those with high school education or less: males, 15%; females, 10%.

TABLE 11.10

High School Integration and Knowledge of Another Job by Age and Sex of Respondent (Percentage Naming Another Employer)

	North, integrated	North, segregated	South segregated
MALES			
Under 30	38 (212)	38 (169)	38 (202)
30-39	46 (195)	10 (68)	30 (189)
40+	40 (106)	15 (27)	24 (152)
FEMALES			
Under 30	35 (306)	34 (174)	21 (248)
30-39	24 (237)	18 (78)	27 (317)
40+	30 (106)	19 (53)	20 (200)

benefit from informal contact. Even the most casual information about employment can be valuable. One irony is that if a single black is hired by a large plant, there are more white employees who know that the firm is integrated than do blacks; thus we arrive at the curious hypothesis that whites will have more information about jobs which are becoming "open" than will blacks.

Social Relations in Integrated and Segregated Schools

This picture of integration as a virtually unmixed blessing seems strange. If anything, it is "obvious" that the student in the integrated school should be the one to drop out, or at least not go to college. There are several reasons why we would expect this. First, we would expect discrimination from white classmates and teachers to make school an unpleasant experience. Secondly, we would expect the black student to make poorer grades than the whites, thus damaging his self-esteem; and third, we would expect the white counselors to either ignore the aspirations of black students or at the least have inadequate knowledge about their college opportunities. After all, these students are not the class of 1971, where black applicants are at a premium; they are the class of 1963, or 1950, or 1939, and we all know, or think we know, what the so-called integrated school was like in those days. These are good reasons not to expect what we have found.

Respondents from integrated high schools were asked: "In some schools, Negro students weren't allowed to participate in some activities—or some classes, like swimming, were segregated. Were the Negro students in your school discriminated against or segregated in the school you went to in activities like these?" Nineteen percent of the respondents said yes—46% of these referring to segregation or discrimination in the use of facilities, and 36% referring to segregated extracurricular activities.

Respondents were asked a number of questions about their high schools and elementary schools. Briefly the pattern is that southern high schools are rated worst in terms of facilities, but teachers and other students are rated more favorably; the respondent is more likely to say that none of the teachers were "mean," for example. Students from southern high schools are also more likely to describe their classmates as studious.[12]

Respondents from integrated schools are more likely to describe their school facilities favorably than respondents from segregated northern schools. They are also more likely to say that their classmates were studious and college-bound. What is surprising is that alumni of integrated schools are just as likely to feel warmly about their teachers as those from segregated schools. For example, 64% of the students from segregated schools say that one of the teachers took a special interest in them, and 58% said a teacher tried to persuade them to go to college. But the percentages of alumni from integrated schools who say that a teacher was interested in them and encouraged them to go to college are equally high—61% and 56%. Apparently black students related as well to the white teachers in integrated schools as they did to teachers in segregated schools, at least some of whom were black. In elementary school, students from integrated schools were clearly more positive about their teachers. For example, 62% said that none of their teachers were mean, and 86% said there was at least one teacher they "liked" in their integrated elementary school. These percentages in segregated schools are 54 and 81%; 8 and 5% lower.

---

[12]These data appear in "School Integration and the Academic Achievement of Negroes," *op. cit.*

So the rather "obvious" argument—that occupying a minority status in an integrated school poses social or psychological strain—receives no support from our data, despite the presence of discriminatory practices in one-fifth of these high schools.

The Psychological Effects of School Integration

We have already demonstrated that school integration develops personality characteristics which promote achievement. As we have seen, integration increases the "security" group of personality attributes—internal control, happiness, and reduced antiwhite sentiment; while decreasing the "assertiveness" factor—expression of anger and self-esteem.

The "contextual" and "interactional" effects are derived from high school integration. But the effects of integration on psychological security are entirely at the elementary school level as Tables 11.11 and 11.12 show. In Table 11.11, we see that northern men from integrated elementary schools have high control, regardless of whether they went to a segregated high school or not; 68% from integrated high schools score high, compared to 70% from segregated schools. And those from segregated elementary schools score low, also regardless of high school (the percents are 59 and 61%). In the other three lines, there is considerable fluctuation, but overall, there is a consistent difference favoring elementary school integration, as shown by the net partial percentage difference at the bottom of the table.

TABLE 11.11
School Integration and Feelings of Control, by Sex (percent with high control)[a]

	Integrated high school		Segregated high school	
	Integrated elementary school	Segregated elementary school	Integrated elementary school	Segregated elementary school
MALES				
Born in North	68 (229)	59 (109)	70 (34)	61 (131)
Moved North before age 10	58 (71)	66 (61)	67 (12)	66 (71)
FEMALES				
Born in North	61 (266)	58 (111)	45 (38)	59 (114)
Moved North before age 10	68 (121)	48 (63)	87 (39)	48 (60)

[a]Net effect, controlling on region of birth, sex, and elementary or high school integration: of elementary integration = 9%; of high school integration = 0%.

TABLE 11.12

Percent Saying They are "Very Happy," by Integration of Elementary and High Schools, Region of Birth, and Sex[a]

	Integrated high school		Segregated high school	
	Integrated elementary school	Segregated elementary school	Integrated elementary school	Segregated elementary school
MALES				
Born in North	34 (229)	10 (111)	12 (34)	15 (131)
Moved to North before age 10	34 (67)	23 (61)	0 (12)	35 (75)
FEMALES				
Born in North	33 (263)	10 (111)	26 (38)	26 (118)
Moved to North before age 10	21 (121)	14 (63)	15 (39)	21 (52)

[a]Net effects, controlling on region, sex, and one type of integration: of elementary integration = 12%; of high school integration = −1%.

The pattern for happiness (Table 11.12) is clearer. In general, the happiest adults are those for whom both elementary and high school were integrated. The next best situation was for both schools to be segregated. If a student attended a segregated high school he was actually worse off if his elementary school was integrated. Finally, the worst combination of all is to come from a segregated elementary school to an integrated high school. (We shall address ourselves to this finding—that consistent schooling is better than inconsistent—in the next section.)

As the summary net effects show, the overall result is that it is only elementary school integration which increases happiness. High school integration, without a foundation of integrated experience in elementary schools, may actually be harmful.

Partial Integration: Marginality and Culture Shock

No one is completely segregated, and no one is completely assimilated. Rather each person finds himself moving back and forth between segregated and integrated experiences throughout his or her life. In the last section we saw that adults who attended a segregated elementary school and an integrated high school, or vice versa, were unhappier than those who attended schools which were either consistently segregated or consistently integrated.

There are two different concepts which might explain this pattern. First is the concept of the "marginal man"—a phrase coined by Everett Stonequist,[13] a student of Robert E. Park. By "marginal," he meant a person on the margin between two cultures, without a clear sense of belonging to either. The half-caste, who actually is ethnically of two cultures, is the perfect example; but the recently promoted shop foreman, who does not know whether he is "management" or "one of the men," is another.

The student who moves from an integrated elementary school to a segregated high school may experience this sense of marginality. The ideal example might be as follows. Last year this student was established in an integrated setting, comfortable in the knowledge that white teachers and white classmates liked him, that he could compete successfully with whites. Now he finds himself in a segregated school. What happened? Why has he been "put back?" Is it true that he is "just a Negro," and the contact with whites in elementary school was merely an accident? In this sense, the transfer student might well experience a loss of identity or self-esteem.

The other concept appropriate here is "culture-shock"—the reaction to being suddenly placed in an alien culture. Supposedly, the person who experiences culture shock reacts by withdrawing, by clinging somewhat desperately to his old culture or his old identity. The long-term effects of this could well produce the unhappiness experienced by the student who moves from a segregated school in the South, to an integrated school in the North. We have already seen that integration in the South has negative consequences. Social contact with white children within a segregationist culture may produce a sense of marginality. We also saw in Table 11.12 that Southerners who migrated to the North and attend integrated schools are as unhappy as migrants who attend segregated schools. This could also be an example of the negative effects of inconsistency in one's racial experience. The child raised partly in the South may find the egalitarianism of the integrated school too frightening. (Of course, his southern accent and the fact that he did not attend kindergarten may handicap him, but these would be serious handicaps in a northern segregated school as well.)

It seems that respondents from the South are socialized into the caste system in such a way that any equal-status contact with whites is disconcerting. This fits with the expression one sometimes hears about the South: "At least in the South you know where you stand."

This means that for southern migrants, the ghetto, as a point-of-entry into the city, may serve a useful function. If it were possible for the second generation to leave and become assimilated, the ghetto would operate much as the ethnic

[13]Everett V. Stonequist (1937), *The Marginal Man,* Scribners, New York; see also Robert E. Park (1950), *Race and Culture,* Chapter 28, Free Press, Glencoe.

neighborhoods did for other immigrant groups: as a place to pause in the process of assimilation. As we have seen, however, this is not the case with blacks.

Inconsistency between neighborhood integration, school integration, and friendships with white children can also have negative effects, just as inconsistency between elementary and high school integration does. For native northern men, inconsistency between integration of neighborhood and elementary school is associated with lower educational attainment, lower verbal achievement, and lower internal control. Since internal control and self-esteem are related in opposite directions to integration, it is particularly interesting that men from inconsistent neighborhood/school combinations have lower internal control *and* lower self-esteem.

The data for self-esteem are given in Table 11.13 for northern-born men. In the upper-right corner are the men with consistently segregated experience: living in a segregated neighborhood, attending a segregated school, and playing only with blacks. As we move across and down, we encounter various combinations of integration and segregation, and self-esteem declines, until we encounter those persons who had two integrated experiences mixed with one segregated experience—either they played with whites and attended integrated schools, but lived in a neighborhood with only a few whites, or else they lived in integrated neighborhoods and played with whites but attended segregated schools. These two categories have the lowest levels of self-esteem. Those whose experience was consistently integrated have slightly higher self-esteem.

Table 11.14 repeats this analysis using internal control, with precisely the same result: those groups which had two integrated experiences and one segregated experience have the lowest control, and those groups with consistent experience, either segregated or integrated, have the highest control. When we recall that internal control and self-esteem are weakly associated, the consistency

TABLE 11.13

Consistency in Integration and Self-Esteem for Northern-Born Men (% with high self-esteem)[a]

Number of whites in neighborhood	Segregated school		Integrated school	
	Did not play with whites	Played with whites	Did not play with whites	Played with whites
None	75 (59)	(64) $(28)^b$	(4)	(10)
Few	(67) $(48)^b$	66 $(113)^b$	(6)	55 $(113)^c$
Many	(5)	53 $(47)^c$	(0)	65 (178)

[a]Association between inconsistency and self-esteem: $\gamma = -.15$; ( ) around percentages based on 20 to 50 cases; percentages not computed for $N < 20$.

[b]Moderately inconsistent statuses.

[c]Very inconsistent statuses.

TABLE 11.14

Consistency in Integration and Internal Control for Northern-Born Men (% with high internal control)[a]

	Segregated school		Integrated school	
Number of whites in neighborhood	Did not play with whites	Played with whites	Did not play with whites	Played with whites
None	45 (55)	(36) (28)[b]	(4)	(10)
Few	(25) (48)[b]	34 (111)[b]	(6)	30 (113)[b]
Many	(5)	30 (53)[c]	(0)	47 (176)

[a]Association between inconsistency and internal control: $\gamma = -.18$. ( ) around percentages based on 20-50; percentage not computed for $N < 20$.

[b]Moderately inconsistent statuses.

[c]Very inconsistent statuses.

between the two tables is impressive. Unfortunately, we cannot conclude from this anything more than that mixed experiences are bad. But since everyone has a mixture of segregated and integrated experience it becomes a matter of degree, and how much of each constitutes inconsistency is impossible to decide. This is also one part of our analysis where considerable change may occur in the short span of two decades, and what was a discomforting experience for our respondents may not be to the present generation.

One of the problems with the marginality hypothesis is that one could read it as predicting that *any* form of integration would have unfortunate psychic consequences. Indeed, Rosenberg[14] finds that Catholics raised in predominantly Protestant neighborhoods have lower self-esteem. Remember that Rosenberg uses a measure of self-esteem which is quite different from ours, and that may be close to what our happiness question is measuring. If that is the case, why is "integration" bad for Catholics and good for blacks? It may be that there is nothing innately depressing about a heterogeneous environment, and the Catholic situation analyzed by Rosenberg may be an exceptional case. But even if in most cases it may be more supportive (if uninteresting) to be surrounded by people who are identical to you, it is easy to argue that the black-white relationship is a special case. Blacks must be permitted to establish personal relationships with whites in order to escape the traditional self-perception of their position as that of a hopeless lower caste. There are a number of caveats. Of course, this contact must be on an equal-status basis; contact between landlord and tenant, or guard and prisoner, merely reveal that the sense of hopelessness is justified. If interracial contact is to serve the purpose of encouraging a black to achieve, the

[14]Morris Rosenberg (1965), *Society and the Adolescent Self-Image,* Princeton Univ. Press, Princeton.

whites must be either unprejudiced or willing to conceal their prejudice. And the black must be "ready" to learn.

Playing with White Children and Anti-white Sentiment: The Contact Hypothesis

Wilner, *et al.* stated the contact hypothesis as follows:

> ... the general hypothesis that equal-status contact between members of initially antagonistic ethnic groups under circumstances not marked by competition for limited goods or by strong social disapproval of intergroup friendliness tends to result in favorable attitude change.[15]

One can probably cite more evidence to support this proposition than almost any other in sociology. It would be completely inexplicable if the contact hypothesis did not also hold for anti-white feelings among blacks, and we shall now demonstrate that it does indeed apply to blacks. Respondents were asked "Did you play with white children when you were growing up?" Of course there is a difference between play in a segregated southern town and play in a northern city, so in Table 11.15 we have separated respondents by sex, region of birth, and integration of neighborhood and school before looking at the effects of playing with whites on anti-white feeling. But the result is unequivocal. In nine of the 12 tests, those respondents who say they played with whites are less anti-white than are respondents who experienced the same migration pattern and degree of neighborhood and school integration but did not play with whites. All three of the exceptions, where playing with whites did not lead to lower anti-white scores, were in the South.

Summary

Integration does have important psychological and social consequences. Overall, integration seems to increase a black's sense of security, while also inhibiting his expression of aggression. Since easy expression of aggression is a mixed blessing anyway, the net result is that integration serves to establish the preconditions for educational and occupational achievement and for a generally happier life. We have seen that respondents who attended integrated schools are better educated, earn more money, and say they are happier.

Beyond this, the effects of integration depend on other factors. The effects on the personality seem to result more from the *consistency* of one's experiences

---

[15] Daniel M. Wilner, Rosabelle Price Walkley, and Stuart W. Cook (1955), *Human Relations in Interracial Housing: A Study of the Contact Hypothesis,* Univ. of Minnesota Press, Minneapolis.

TABLE 11.15

Playing with Whites and Anti-white Feeling by Sex, Region of Birth, Neighborhood and Elementary School Integration (% low in anti-white feeling)[a]

Region of birth and neighborhood	School	Men		Women	
		Did not play with whites	Played with whites	Did not play with whites	Played with whites
Northern-born					
Black	Segregated	69 (55)	86 (28)	88 (82)	95 (39)
Black	Integrated				
Few whites	Segregated	58 (48)	84 (112)	71 (34)	89 (106)
Few whites	Integrated		78 (113)		84 (109)
Many whites	Segregated		58 (53)		91 (32)
Many whites	Integrated		78 (176)		83 (192)
Moved North before age 10					
Black	Segregated	89 (27)		61 (28)	100 (23)
Black	Integrated				
Few whites	Segregated	45 (40)	70 (54)	67 (24)	77 (65)
Few whites	Integrated		88 (34)		86 (59)
Many whites	Segregated		77 (26)		
Many whites	Integrated		90 (31)		80 (71)
Moved North after age 10					
Black	Segregated	72 (247)	70 (96)	76 (314)	72 (89)
Few whites	Segregated	58 (148)	68 (333)	75 (281)	78 (342)
Many whites	Segregated		85 (91)	72 (47)	65 (106)

[a]Percentages not computed for N < 20. Case bases omitted for clarity.

with integration than from any specific instances of integration. Part of the explanation for this may be that the personality develops over time, with certain phases of one's life being more important than others to the development of certain traits and skills. For example, elementary school integration has an important effect on personality, as we have shown, but southern migrant children tend to be unprepared for integration and therefore cannot benefit at this level. Expectations learned in the South are unfulfilled by the integration experience, leaving the individual disoriented.

The effects of integration which proceed directly from interaction with whites (e.g., obtaining job-finding information, adopting a norm regarding finishing high school or going to college) occur at a later age, so that both migrants and Northerners benefit from the experience if they have been prepared for it.

Finally, the whole case is complicated by the change that has occurred in American society since our respondents were growing up. For example, it may no longer be true that integration inhibits aggressiveness. Racial tension in integrated high schools is hardly new, but it may be growing in severity; if it is, this may be a symptom that inhibition is less common now. Furthermore, it may no longer be true that the South is more oppressive than the North, although the same principle of consistency would still apply.

But we do not think that 30 years is long enough to change the fundamentals of the process we are dealing with. In many ways what is important about school integration now was also important when suit was filed against a segregated Boston school system in 1837.

12
The Policy Issues: An Essay

It seems to us that the evidence is convincing for the King-Kardiner hypothesis. Discrimination does lead to psychological disability which in turn prevents achievement. In trying to ease the transition from slavery to full equality, we have made a disasterous wrong turn, creating in the northern ghettos a tinderbox of pessimism and alienation. At this point the simple truth of segregation seems to be this: segregation serves as a highly visible symbol of white racism, making blacks furious and hopeless, and unable to function in their fury and hopelessness as fully adequate men and women.

Stated this way, it seems obvious, but it is also "obvious" that the world is flat. More important, the first statement is true. What parents tell their children about race is more important to a southern black than whether he comes from a broken home; to a northern black, whether he attends an integrated elementary school is more important than whether he comes from a culturally deprived home. For that matter, it now seems likely that his reaction to being black may be a critical factor in determining whether his children grow up in a broken or culturally deprived home.

Our data analysis ends with this conclusion and it would be appropriate to end the book at that point. This additional chapter goes "beyond the data" to consider the implications of these findings. In his Presidential address to the American Sociological Association, Everett C. Hughes called his fellow social scientists to task for making no effort to anticipate the future of race relations.[1]

---

[1] Everett C. Hughes (Dec. 1963), "Race Relations and the Sociological Imagination," *Amer. Soc. Rev., 28,* 897–890.

Earlier, Robert Lynd appealed to sociologists to put their science to work for humanity, in a book with the eloquent title "Knowledge for What?"[2]

We shall take these two appeals seriously in this last chapter, and attempt to make some projections regarding the future of the ghetto and consider some of the ways in which society might intervene to alter that future.

Projections for the Future

If a series of alternative projections are posed, ranging from very bleak to very optimistic, it seems safe to reject as unlikely the bleakest possibility: that of widespread racial conflict in the form of riots. The riots were a form of "fad," spreading rapidly across the United States and dying out just as quickly.

The recent guerilla warfare between the police and some "lunatic fringe" groups such as the Panthers may also be a fad, but if it is, it may die much more slowly. The riots had a transparently self-destructive quality which probably worked to deter their recurrence; but murdering policemen from ambush is not so obviously hazardous. And violence does meet a real need to express anger toward whites, to release aggression, and perhaps to prove one's manliness. The ghetto has always had a high level of violence, mostly directed at other blacks. Anti-white violence, whether in the highly visible form of guerilla warfare or the more subtle form of assaults on individual whites, can be expected to continue indefinitely.

The New Militancy

We can also expect at least one aspect of the racial situation to change: blacks should become more ethnocentric. The nation has been fortunate that up until very recently black leaders have stressed integration and assimilation. In the past, anti-integrationist leaders such as Marcus Garvey have been held in disrepute by black elites. This happened for two reasons: first, the assimilation model is the only one which seemed to have any realistic hope for success; and second, the, absence of a distinct black culture plus the inhibition of anti-white feeling meant there was little support among the masses for a revolutionary or separatist ideology. But ethnocentrism is an almost ubiquitous characteristic of humanity, and as the legal and social status of blacks increases, this ethnocentrism should become more apparent. Thus while some aspects of the militancy of the 1968-1970 period may be fads, separatism and anti-white hostility itself should remain indefinitely.

There is not much reason to be concerned about ethnocentricity *per se.* Nor is black militancy necessarily dysfunctional. To the extent that militancy means

[2] Robert S. Lynd (1939), *Knowledge for What?* Princeton Univ. Press, Princeton.

the aggressive pursuit of legal and economic opportunities and judicious use of bloc voting, it will almost certainly benefit blacks. Ethnocentricity and militancy only become dysfunctional when they involve a blanket anti-white hostility, or a cover for segregationist policies. The new militancy of the 1970's does have definite elements of both in it.

There is nothing in our data to suggest that militancy is conducive to achievement. As we pointed out earlier, militancy does not seem to be a "therapy"—a way of changing behavior so as to increase achievement. Rather it seems a rationalization of the low internal control/high self-esteem mood which has prevented black achievement in the North.

If expression of anti-white sentiment was therapeutic, then a high score on the antiwhite scale would not be associated with a low sense of internal control and with unhappiness. One might argue that this pattern is changing in the younger generation. But in fact it is not. The association between anti-white attitudes and internal control remains negative ($\gamma = -.42$ for men under age 30 compared to $\gamma = -.48$ for older men). Young high school graduates have a lower sense of internal control than high school graduates over age 30.

Lee Rainwater points out that white racism has always been able to depend on the unwilling cooperation of blacks in maintaining a lower caste position for blacks.[3] In the past, white racists have benefitted from Uncle Tom, and blacks have suffered from the internalized and deflected aggression. Now white racists are benefitting from the black segregationists' defense of the status quo; and while in the past, whites were the primary teachers of the doctrine of black hopelessness, black militants are now preaching some elements of the same theme.

The low sense of internal control among young blacks is discouraging. What are the prospects that this trend will reverse? Objectively, black life-chances have improved; there are more opportunities in certain high status occupations, and the percentage of blacks finishing high school has increased very rapidly. The problem is that these changes must be *perceived* by blacks as meaning improvement. It is very difficult to prove that a particular institution discriminates against blacks, but it is even more difficult to prove the absence of discrimination. The only hope is for blacks to believe that there is less discrimination, that their chances are better. At the same time that the objective situation is improving, blacks are freeing themselves from the inhibitions that acted in the past to prevent them from seeing discrimination. Thus objective improvement may be occurring simultaneously with subjective decline in optimism.

The "black pride" movement, which had just begun when this survey was done, may have considerable impact. Pride in blackness has already apparently destroyed the favoritism blacks had previously shown to light skin. As the next generation grows up in a climate of the new black consciousness, it may arrive

[3] A rough paraphrase of Chapter 1 of Lee Rainwater (1971), *Behind Ghetto Walls,* Aldine, Chicago.

at a new sense of identity which is more conducive to achievement. It certainly seems reasonable that holding Ben Banneker and Harriett Tubman up as role-models for young blacks should make a difference.

Prospects for Government Intervention

We noted in Chapter 3 that the poverty program seemed based on the theory that poor blacks were not different from poor whites. We now see that this is an inadequate theory, and there is little reason to expect it to lead to sound policy planning. The problems of black poverty are problems partially caused by poverty, and to this extent the poverty program will be effective, but equally important causes are discrimination, marital instability, aggression, and hopelessness, and the melange of federal programs now in operation does little in any of these four problem areas. Headstart, the Job Corps, and job-training programs do not address the fundamental issues.

The various antipoverty projects now underway are valuable; their results can be seen in the increase in black family incomes and the large decrease in school dropout rates during the late 1960's.

Our data indicate that welfare, employment, and education programs will have positive effects on the black personality. Families with incomes over $4500 are happier, for example, and it seems safe to assume that children of happy parents will achieve more. Furthermore, this is one area where the society may take action; there is good reason to believe that welfare payments will increase during the 1970's, and federally funded day-care centers are likely.

Finally, we can expect the government to do more to break down discrimination in employment. The federal government has recognized this as a legitimate national goal, and the results are already being felt in the gain of black income relative to whites. But much more must be done. It goes without saying that there is no point in asking blacks to lift themselves up unless the "job ceiling" is removed.[4]

We should also recognize that not all aspects of the black-white personality differences are dysfunctional. The low level of verbal aggressiveness which adult blacks show is offensive because it seems appropriate to the servile role of the bootblack or chauffeur. The same verbal style might be applauded from a black policeman or mayor. To cite two examples: the long years of second-class citizenship have apparently taught blacks to empathize with the underdog and to support collective action to help the poor. The result is that black voters can be counted on to support higher taxes for education and social welfare much more than do whites of the same income.

[4] The phrase is from St. Claire Drake and Horace Cayton (1962), *Black Metropolis* (paperback ed.), Harper, New York.

Second, Martin Luther King, Jr., was one of the first political figures to criticize the Vietnam War. He thus provided a useful service to the nation; for no matter how the reader feels about that war, he will probably agree in wanting to get it over with. Many political observers assumed that King had blundered in going outside the area of civil rights, and that he would lose his political support among blacks. These commentators did not realize that Christian pacifism is valued more strongly by blacks than by whites—at least, this has been found in surveys since the beginning of World War II.[5] Empathy, support for government action, and pacifism are as much a part of black personality structure as are the dysfunctional traits we have focused upon in this book.

On the whole, unfortunately, we are not at all optimistic. Without a radical change in race relations we can only expect a continuation of the present trend: a slow growth in income and rapid increases in levels of education, coupled with continued or increasingly high levels of marital instability, violence, and alienation. Thus it seems that integration begun in childhood will be necessary before we can hope for dramatic change in black achievement. This means the first priority should be given to a major program of integration in schools and housing.

School Integration

While the federal government has acted to bring integration to many southern schools, much less has been done about integration in the North. Most northern schools have done a token amount to create integration, but only a few systems have gone beyond this. A number of different arguments against integration have been presented: (1) integration is impossible because the large central-city school systems are already overwhelmingly black; (2) blacks no longer want integration; (3) one cannot change attitudes; whites are opposed to integration, and will refuse to permit their children to be bused; (4) the neighborhood school is educationally better than busing; (5) desegregation efforts should be directed against housing, which is the basis of school segregation.

When these objections are coupled with the fact that many school administrators still see integration as a "social" rather than "educational" issue, integration stands little chance. It is interesting that the Coleman Report's findings have not influenced very many administrators to see integration as an educational tool. Perhaps this is because there has been no theoretical basis for expecting integration to lead to achievement. Theories of "cultural deprivation" and even genetic differences are perhaps easier to accept.

It is of course true that poor black children do live in a culturally deprived environment which hampers their school achievement. But the school is part of that environment, and school integration is a simple and effective method of cultural enrichment.

[5] Personal communication to Robert Crain from Sidney Verba.

While genetic theories may be intellectually interesting, there is only meager evidence, and the policy-maker should not be too impressed when he compares this to the empirical evidence supporting the case for school integration.[6]

While the arguments against integration must be considered, they do not mean that integration is impossibly difficult for the school system which wishes to develop an effective plan. It is true that all the schools in Washington, D.C., cannot be integrated, except by making all the schools have an equal 95:5 black-white ratio. But the Chicago public schools are only 52% black, and there is no mathematical reason why only 3% of the blacks in Chicago are in predominantly white schools.[7]

If we bused children from the ghetto into white schools with travel time not to exceed 30 minutes, and added portable classrooms so that each receiving school could increase its total enrollment until it was one-third black, it would be possible for over 25% of Chicago's black students to be in predominantly white schools, an eightfold increase over the present situation. Applying the same techniques in Milwaukee, it would probably be possible to integrate all the black students in the city. (At present 12% are in predominantly white schools.)[8]

It is true that there is less demand for integration from civil rights groups now than 5 years ago, but this does not mean that black parents would refuse the opportunity to send their children to integrated schools. In our survey black parents put education in general as a high priority civil rights goal, but seem to consider school integration specifically as much less important. Thus it seems likely that black parents have not changed their views very much; they still view school integration as a good thing, but not of great importance. The new militancy does not mean that black parents have become opposed to integration.

Experience and survey data both indicate that whites will usually not object to black students being bused into "their" schools.[9] However, one can expect

[6] The "genetics" and "integration" arguments are not incompatable. No one contends that scholastic achievement is entirely genetic, and even accepting the calculations of Arthur Jensen, there is still room for environmental factors (such integration) to make a considerable difference in intelligence. See Arthur R. Jensen (Winter 1969), "How Much Can We Boost IQ and Scholastic Achievement?" *Harvard Educ. Rev.,* 32, 1-123. A series of replies to the Jensen article were published in that journal over the following two years.

[7] Data from Department of Health, Education and Welfare, News Release dated January 4, 1970; raw data published in *Directory of Public Elementary and Secondary Schools in Selected Districts,* Fall 1968, Health, Education and Welfare, Office for Civil Rights, Washington, D.C.

[8] The limitations would lie in the number of black students who would object to being bused, the problems of disposing of ghetto school buildings as they are emptied, and financing additional construction in white areas.

[9] See Robert L. Crain with the assistance of Morton Inger, Gerald A. McWorter, and James J. Vanecko (1969), *The Politics of School Desegregation,* NORC Monographs in Social Research, Aldine, Chicago. See also Paul B. Sheatsley (1966), "White Attitudes Toward the Negro," *Daedalus,* The Negro American, 2, (Winter) 95, 217-237.

considerable opposition to a "reverse busing" scheme of sending whites into ghetto schools. New York City, after several years of busing black students into white schools, encountered significant opposition when it attempted to send white students to predominantly black schools in its abortive plan to "pair" white and black schools.

In summary, the most successful integration plan would be one which avoided creating predominantly black schools (the educational gain for blacks is less and the white opposition greater) and avoided sending white children into ghetto schools (where problems of violence might be more serious and white parents would object). This means that the basic technique would be to bus from the ghetto to white areas, integrating as many different white schools as possible and using portable classrooms and other devices to provide the needed space.

It may be that busing would provide an uncomfortably inconsistent interracial experience for some blacks. Our data will not give us definite answers on this point, partly because our data deal with the effects of integration 15 or more years ago. It is probably true that busing is psychologically less satisfactory than living in an integrated neighborhood; it is also likely that today's generation of black children will be less uncomfortable when bused into a white school than their parents would have been. Even if busing has some negative consequences, these are probably outweighed by the various benefits from integration which the bused child would receive. And the school administrator interested in integration can hardly afford to wait until that time when housing is integrated; busing is the only practical alternative at this time.

School integration represents a partial solution to the problems of black poverty. But what are the prospects that it will happen? A few northern school boards in smaller cities have taken action. But there is no reason to expect the average school system to do so. School integration is too controversial, too innovative, too radical a departure from traditional teaching philosophy. More important, school integration lacks a constituency to bring action out of school systems which are normally more reactive than initiating. A sustained drive by local civil rights movements would probably show some success, but there is no reason to expect the civil rights movement to reappear, or to return to school integration. Even in cities which have elected black mayors, school integration has not received high priority.

There is not much reason to expect the federal government to force school integration. It is one thing to bring the power of the government to bear on a minority region in a "second reconstruction," it is less likely that the government will act against the apparent wishes of a northern white majority. There is, however, the possibility of Supreme Court decisions eventually forcing limited integration.

Finally, there is some possibility that Congress might adopt a "carrot" rather than "stick" approach, offering funds to aid in desegregation. The federal government could simply make its federal aid in the form of sizable subsidies to local districts for each student in an integrated school. If the federal government were to key federal aid to education to a performance criterion—if school systems received aid proportional to gains in achievement, just as Kenneth Clark and others have proposed that teachers' salaries be keyed to the achievement of their pupils—then a few far-sighted systems might integrate schools in order to push test scores up. At this point, such a thought seems to border on social science fiction.

In conclusion, we can expect a steady but slow increase in school integration, but far below the maximum amount that could be achieved. There is reason to be hopeful, in that largescale school integration presents no technically insoluble problems in the implementation of a desegregation plan. The problem is that very few school systems show any inclination to do so. There are a number of different ways in which federal and state governments might act to encourage local systems to integrate. Unfortunately, none of the possible political solutions to the dilemma seems at all likely at this time.

Housing: A Brighter Forecast[10]

It may seem paradoxical that we can expect more success in integrating housing—where nothing has been done so far—than in schools, where integration has been occurring in the North since the early 1960's.

One reason why desegregation could begin in the schools is that they are highly centralized; there was a focal point at which to apply pressure. But this is also the reason why schools will be difficult to desegregate totally—for action must come from the top. If a family decides to send a child to an integrated school, the system must provide a seat in an integrated school, permission to go there, and a means of transportation. Furthermore it must overcome the obvious fact that for the parent, it is easier and more convenient to send the child to the neighborhood school. No doubt it would be possible to persuade a large number of black parents that it is to their advantage to send their children out of the neighborhood, but we know of no northern school boards which are willing to encourage black children to transfer to integrated schools.

In contrast, there is no center to the housing market; no one button to push. But at the same time it is so obviously to the advantage of a black family to buy

---

[10] A great deal has been written on housing integration. The interested reader should see Norman M. Bradburn, Seymour Sudman, and Galen Gockel (1970), *Racial Integration in American Neighborhoods*, The National Opinion Research Center, Chicago, (which contains a bibliography), and Anthony Downs (1970), "Alternative Futures for the Ghetto," *Urban Problems and Prospects,* Markham, Chicago. For white attitudes toward integrated housing, see Angus Campbell (1971), *White Attitudes Toward Black People,* Univ. of Michigan, Institute for Social Research, Ann Arbor, Michigan.

housing in integrated areas, that one need only find a means to permit them to do so, and each individual family will become a force for integration. Just as there is no one button we can press to start integration, there is no one button which can prevent it.

Let us explain what we mean. Taken at its most abstract, housing segregation is a collective decision by the more powerful race to withhold a scarce resource, land, from the weaker race. If the minority population is growing, the demand for housing increases but the supply is restricted. Thus blacks pay higher prices for land and housing than do whites. Eventually, these prices become sufficiently high that profiteers—"blockbusters"—will purchase some housing and transfer it to black ownership, profiting from the difference between black and white market rates.

Segregation in housing depends upon the tacit agreement between white home-owners and white realtors to refuse to sell to blacks. But each home-owner and realtor, by voluntarily restricting his market, is sacrificing money. It is no more profitable for a home-owner to segregate than it is for a corner druggist to refuse to serve black customers; both are simply limiting their market. Further, since integrated housing is less expensive than comparable black housing, black home-buyers have an important incentive.

This means that segregation in the housing market is a house of cards; once it begins to collapse, it will fall of its own weight. This does not mean that it in fact will fall. After all, it is obviously a waste of money for men to wear ties or women to change skirt lengths, but this in itself does not mean it will not be done.

There are usually three arguments made against efforts to desegregate housing: (1) blacks cannot afford to live in the suburbs; (2) blacks do not want to integrate; and (3) whites will lose money.

The answer to the first argument is an empirical one and has been studied by Karl and Alma Taeuber.[11] They compute an index of housing segregation, by examining the statistics for city blocks from the census, and computing the fraction of all blacks who would have to move in order for each block in the city to have an equal number of each race. For example, the figure for Detroit in 1960 was .79. Seventy-nine percent of all blacks would have to be relocated. To determine how much of this segregation could be attributed to differences in income, they computed the theoretical index of segregation for the situation in which blacks and whites were segregated only because of differences in quality or value of housing; this figure for Detroit was 17.5%. Much more segregation was the result of race than of income. Results for other cities were similar.

The question of whether blacks are willing to move into integrated neighborhoods is complex; what people say they will do in the absence of opportunity is

[11] Karl Taeuber and Alma Taeuber (1965), *Negroes in Cities,* Aldine, Chicago.

not necessarily what they *will* do. In our survey, we asked, "Do you think you and your family would prefer to live in a neighborhood which was mostly white, one which was integrated but mostly Negro, or one which was all Negro?" Twenty-two percent said mostly white, 58% mostly black, and 9% all black. Angus Campbell and Howard Schuman[12] worded the question slightly differently for their survey: "Would you personally prefer to live in a neighborhood with all Negroes, mostly Negroes, mostly whites, or a neighborhood that is mixed half and half?" Thirty-seven percent said it makes no difference, 48% said half-and-half, 8% said all black, 5% said mostly black, and 1% said mostly white. The Campbell and Schuman survey is of 15 large city ghettos. These responses indicate a strong commitment to the principle of integration, but relatively little willingness to be part of a small minority.

In another question we asked:

> Suppose someone came to you and told you that you could rent or buy a nice house that you could afford, but it was in an all-white neighborhood and you might have some trouble out there. Are you the pioneering type who would move into a difficult situation like that?

Forty-five percent said yes. We also asked "Have you ever tried to find a house or an apartment in a neighborhood which was mostly white or all white?" and 28% said yes to this. Together, this suggests that as long as better housing is available in white areas, blacks will be willing to live there. The question which cannot be answered is whether blacks will show a distinct preference for integrated areas where there are many blacks. In one sense this could be expected; most cities have ethnic suburbs of middle-class housing. However so many factors go into the housing decision that most families do not choose a house on purely ethnic grounds.

The Question of Property Values

The final question—whether white property values will decline or not—has both a simple answer and a more complicated one.[13] The simple answer is no; the addition of black buyers to the white market can only increase demand and push prices up.

Again working with a simple theoretical model, we can deduce the following idealized portrait of a changing neighborhood. When an all-white neighborhood

[12] Angus Campbell and Howard Schuman (1968), "Racial Attitudes in Fifteen American Cities" in *Supplemental Studies for the National Advisory Commission on Civil Disorders,* p. 15, U.S. Government Printing Office, Washington, D.C.

[13] A series of studies have been made of this question. See, for example Luigi Laurenti (1961), *Property Values and Race,* Univ. of California Press, Berkeley.

changes, the first person to sell to blacks—the blockbuster—will profit by increasing his price to equal those for equivalent housing in the ghetto. Panic may occur, with a large number of houses on the market, and sellers will receive lower prices. But in the long run, blacks pay more than whites for housing, and most whites profit from selling to blacks.

Actually, housing prices may not be the most important economic factor for white families.[14] Many families can only move at certain points in their life cycle; for example women and older men have difficulty obtaining mortgages. And even if a family received over the "fair market value" of their house, they must find replacement housing. In many cases, families will buy newer, more expensive replacement housing and be somewhat justified in feeling that they have suffered a loss.

Thus far we have discussed changing neighborhoods. Now let us consider what happens when an entire suburban ring becomes integrated. First of all, panic is much less likely. If all housing is integrated, then there is no place for panic-stricken whites to flee. Further, if a number of other areas are also open to blacks, black demand for housing in this area may be very low. Under these conditions, panic will not occur, housing prices will increase slightly, housing prices in the ghetto will drop, and the excess demand created by black buyers will filter up to create an increase in new housing starts. Under these circumstances, there is only one possible condition under which whites would lose money. If one particular area becomes popular with black home-buyers, the area may be labeled as undesirable by whites. The withdrawal of white demand from the area, coupled with only a slight increase in black demand, could depress values in this area. But these cases may be relatively rare since, with many parts of the suburban ring open to blacks, there will not be enough black home-buyers to overload very many areas.

How to Make Integration Work

This argument permits us to develop some recommendations about how to integrate suburban areas:

1. Integration should begin with new housing, higher income housing, and housing farthest from the central city. These are the areas least likely to panic and are also the areas where it will be easiest to achieve integration, since the new home market is in the hands of a small number of builders who can be pressured to conform to the law more easily than individual home-owners. Integration of these areas will prevent panic in other areas; for if one cannot escape blacks by moving to better housing, one cannot panic.

[14]See Anthony Downs, "Losses Imposed on Urban Households by Uncompensated Highway and Renewal Costs" in *Urban Problems and Prospects, op. cit.*

2. The second phase of integration should concentrate on token integration of the remaining white areas, working progressively in from the outskirts of the urbanized area, ending by integrating low-income white areas and areas adjoining black neighborhoods last.

3. The third phase should concentrate on "filling in" these neighborhoods until the black demand for suburban housing is met. Presumably at some point in this phase, natural market forces will take over so that the process will operate without intervention from government.

4. After there is widespread integration of suburban areas, political attention can be turned to the provision of low-income housing.

It is probable that over a long period of time, the buying power of blacks and the profit motive will lead to the collapse of housing segregation. But the process could be speeded up by government intervention. Certainly it should be much easier to integrate northern subdivisions that it was to desegregate southern schools, a job which was done with considerable success.[15]

It seems likely that desegregation by as few as a thousand black families in any one metropolitan area might be sufficient to trigger the opening of housing opportunities for other blacks. As we said, the fact that segregation is economically unprofitable to whites (except for ghetto landlords, who as a class have few friends and are not a political force) means that the present system is unstable.

What actions would be necessary to bring about suburban integration? There are two major obstacles which must be overcome. First, black families who are interested in housing must be informed that housing is available and permitted to "shop" widely, just as whites do. This is a problem which is less critical in school integration, since it is easier to choose between a small number of schools. (This suggests that it will be easier to integrate large housing projects and apartment complexes than older, more diverse neighborhoods.) This also means that integration will not occur simply because of the existence of fair-housing laws, which leave the enormous burden of initiating all the action for the home-buyer.

No integration program will succeed without the cooperation of realtors, at least those realtors presently handling black customers.

[15]At the present time, southern schools are more integrated than those of the northern metropolitan areas. For example while only 3% of the black children in Chicago are in predominantly white schools, 22% of the black students in Greensboro, North Carolina, were in predominantly white schools in 1968. Consider that Little Rock Central High School, integrated by soldiers with fixed bayonets under the command of General Edwin Walker, was 33% black in 1968. (Figures from *Directory of Elementary and Secondary Schools, Fall 1968, op. cit.*) By 1970, the large southern systems were less segregated than those in the North, and this means that the South has changed drastically in a very short period of time. By 1970, 38% of southern black students were in predominantly white schools, compared to 28% in the North. (*Baltimore Sun,* January 15, 1971, p. 1.)

One problem for the black real estate firm is that if a large fraction of white home-owners refuse to sell and must be sued, then the cost of locating sellers in white areas will go up considerably. The second obstacle is anti-black violence from a fraction of the white community. This problem can be addressed in part by integrating high-income areas first. Lower-income white areas would then be more likely to view integration as a *fait accompli* which cannot be combatted. But this is not a total solution. The problem is that a very small minority of rabble-rousers in the midst of a group of whites who are neither for nor against integration can intimidate a neighborhood. In the long run, these problems will "work themselves out." As housing integration becomes common, it will excite no more feeling than integration in employment does now.

Assuming that no Congressional action will be taken to further integration, what other possibilities are there?

Downs summarizes present government policy succinctly: "Not one single significant program of any federal, state, or local government is aimed at altering this tendency (for nonwhites to be excluded from suburbs) or is likely to have the unintended effect of doing so."[16] Given that this is the situation and that Congress shows no inclination to take action, what hope is there? Downs takes a very pessimistic view; he argues that a massive change in public opinion is a political necessity. While Downs' view is not one to be discarded lightly, it seems unduly negative. One possibility is for one federal agency to act on its own initiative, just as the Supreme Court did in desegregating schools, or the Army did in desegregating off-base housing. Neither agency was acting with the consent of Congress. Similarly, the Department of Housing and Urban Development or the Department of Health, Education, and Welfare might decide to act on its own administrative capacity to force desegregation. Such action might take any number of forms; it might be decreed that segregated areas would be ineligible as possible sites for new defense industries (the issue of housing segregation was raised in connection with the location of a major nuclear reactor a few years ago). This step would force a suburb to fill a quota of black housing before becoming eligible, and would force real-estate agents and home-sellers to seek out black applicants.

It is also possible for relocation agencies to take a more active role in locating integrated housing for families who are in the path of expressways, for example. Again, such action might be taken without the explicit approval of either local or federal government.

A third possibility is that large employers—particularly downtown merchants, utilities, and city governments, all of whom would benefit from dispersal of the ghetto—will become involved in counseling black employees regarding housing opportunities.

[16]Downs, *op. cit.,* p. 29.

Unfortunately, neither the federal nor local government has shown any inclination to pursue housing desegregation. While schools could be desegregated through judicial action, the highly decentralized housing market would require innumerable lawsuits. In the case of southern school desegregation, the NAACP was able to achieve considerable desegregation through the courts and this no doubt influenced the federal administration to act. There is no easy way for a group like the NAACP to achieve similar results in housing.

At the same time, local and state governments cannot be expected to act. Central city mayors feel that integration of the city is unrealistic unless the suburbs act also; and the mayor who is willing to exert pressure of this sort on the suburbs surrounding his city is rare. The central city government whose tax base is crippled by its concentration of low-income blacks does stand to benefit from desegregating the suburbs. Downtown merchants, whose white customers are fleeing the city, will also benefit from desegregation. The downtown business interests and city housing departments would also benefit in the long run from the decreased demand for ghetto housing, which would permit demolition of slum housing and expansion of urban renewal. Unfortunately, neither downtown businessmen nor city governments have shown much interest in innovative solutions to their problems in the past, and there is no reason to expect them to do so now.

Integration and Poverty

A very persuasive argument can be made in favor of subsidized housing for the poor in suburban areas. We have resisted focusing upon this strategy simply because it seems politically much more difficult than integration of unsubsidized housing. To build public housing or provide some other form of low-income housing would require a firm commitment from the federal government, including congressional action, which seems unlikely. Under present laws, it would also require the consent of the local government which is even more unlikely. There is nothing technically complex about the necessary legislation; for example, a simple ruling that cities must provide a certain amount of public housing before they can qualify for federal aid would suffice. But to concentrate on this program at this time would be mere utopianism.

It should not be assumed that integration of the suburbs means restricting integration to upper-middle class blacks. While new housing in the suburbs is expensive, there are large numbers of less-expensive older housing units in suburban areas. And families who can afford housing in this price range are not so well-off that we can ignore them. The family with an above-average income, if kept in the ghetto, may produce children with delinquency problems, poor educational opportunities, with less opportunities for teenage employment, less chance to finish high school or go to college.

To put it another way, if we could guarantee that every family which had raised its income above the poverty level would never return to poverty later, the problems of black poverty would be largely solved.

As a program to alleviate poverty, housing and school desegregation cannot reach all of the ghetto residents. Probably less than one-third of ghetto residents could be expected to move into integrated areas within 20 years after housing desegregation occurs; and in the largest cities, only a fraction of those remaining in the ghetto could be bused out to integrated schools. Of course, for a city like Washington, D.C., it would take a number of years to bring large numbers of whites back into the District school system.

Does integration offer anything to those who remain in a segregated situation? It probably does, although there is no way of knowing with certainty at this time. First, the presence of an opportunity to be "successful" by moving to the suburbs should provide a definite incentive. The black husband who wonders what the pay-off is for hard work has a new answer; he can escape the most obvious symbol of second-class citizenship. To cite one example of supporting evidence, the first riot occurred in Los Angeles immediately after a state-wide referendum rejected the right of blacks to live where they chose. There is no way to prove, in this case, that this was a major precipitating factor; but it would certainly be consistent with the general argument presented in this study.

It is true that the poor family is still discriminated against by being made to live in segregated housing, but as King points out in the passage quoted in Chapter 3,

> If one is rejected because he is low on the economic ladder, he can at least dream of the day that he will rise from his dungeon of economic deprivation.

A good case can be made that widespread integration would bring hope to those who remain segregated. And hope is perhaps the one most important thing which ghetto residents need.

There are other, less important, side effects of integration for ghetto residents. One is that the decreased demand for ghetto housing will lower rents in the ghetto—an important factor in a few cities where the "black tax" on housing is high.[17]

In addition, there would be a decreased demand for jobs in the ghetto. There would also be a decreased supply, since some service industries serving the ghetto would lose customers; but the number of potential employees in the ghetto would decrease faster. Desegregated black teenagers would be clerking in suburban stores, thereby reducing the competition for jobs in ghetto stores and downtown shops, for example.

[17] For an estimate of the "black tax" for Chicago see Otis Dudley Duncan and Beverly Duncan (1957), *The Negro Population of Chicago,* Univ. of Chicago Press, Chicago.

Third (and important only in the long run), the decreased ghetto population would increase vacancy rates, thereby permitting government to expand its urban renewal program and construct more ghetto public housing without being stymied by unsolvable relocation problems.

Finally, the presence of integrated neighborhoods will gradually liberalize the views of the whites who come into contact with their new neighbors. More immediately, the presence of a small but solid block of black votes will affect the view of elected officials in some districts where there is close competition for office.

If the federal government committed itself forcefully to integration in both schools and housing, there would be an immediate boost in black morale. The most visible and dangerous aspects of the race problem—racial tensions and violence—might show noticeable improvement within this decade. This means that black poverty would also decline, although the total effects of integration might take two or three decades to appear. Two or three decades is a long time. But given the present state of the society and the conditions of the ghetto, a prediction that things will have improved significantly by the year 2000 might be considered an optimistic one.

The Sampling Design and Sampling Bias

The sample is a "block-quota" sample; in each of the 297 sampling points in the 25 metropolitan areas, a quota for men, working women, and nonworking women aged 21 to 45 was derived based on 1960 census tract statistics.

The design called for 1782 interviews. However only 1651 were completed. Partly this was because the census tract data was out of date and based on nonwhites, rather than blacks; in some tracts there were no blacks to be interviewed.

The sample was stratified by income and size of metropolitan area. Tracts above the median for nonwhite income in the city, and tracts in metropolitan areas of under 2,000,000 population in 1960, were both oversampled by a factor of 2. Then the tape was weighted back to representativeness so that each case was assigned the weights shown in the following tabulation.

	Weight
High-income tracts in small metropolitan areas	1
High-income tracts in large metropolitan areas	2
Low-income tracts in small metropolitan areas	2
Low-income tracts in large metropolitan areas	4

195

The mean weight for the entire sample was 2.52. Block quota samples are less expensive than probability samples, where the names of all residents are "listed" and a random sample selected before interviewing, but usually shows a slight tendency to over represent high-status people. Since the status of northern blacks changed greatly between 1960 and 1966, it is difficult to estimate exactly the size of this bias, but the following calculations give a rough estimate.

There is a large difference between the educational attainment of our sample and that of the northern nonwhite population aged 22-45 in the 1960 Census (Table A1.1). However, most of this difference can be attributed to the increase in education among blacks. If we project the 1960 population forward, assuming persons aged 27 or less in 1966 have the same educational attainment as persons aged 22-24 in 1960, we find that the percent of women who have 8 or fewer years of schooling decreases from 26.6% in 1960 to 21.8% in 1966. The decrease for men is from 31.8% to 26.7%. Table A1.2 shows the differences in educational attainment by age groups in the 1960 census.

TABLE A1.1

Educational Attainment of Nonwhites, by Sex, for Ages 22-45 in the North, 1960[a]

Years of schooling	Males		Females	
	Census	Sample	Census	Sample
0-8	31.8	21.6	26.6	15.0
9-11	29.0	33.0	30.1	38.2
12	24.6	26.4	30.6	30.2
13-15	8.7	14.0	8.2	11.1
16+	6.1	5.2	4.6	5.4
N		(1863)		(2200)

[a]Source: U.S. Census, Special Report, *Educational Attainment,* Table 2, pp. 15-40.

TABLE A1.2

% of Northern Nonwhites with up to 8 Years of Education, 1960

Age	Males	Females	Age in 1966
22–24	17.6	14.3	28-30
25–29	22.8	18.3	31-35
30–34	30.6	24.9	36-40
35–44	40.8	35.7	41-50
Total aged 22–45:	31.8	26.6	

The projection method used above understates the amount of change, since the persons under age 28 in 1966 will have higher educational attainment than those aged 22-24 in 1960. In addition the census data are for nonwhites ages 22-45; omission of 21-year-olds and inclusion of nonblacks affect the educational attainment slightly. Considering these factors, we conclude that persons with elementary education are underrepresented by 3 to 5% in the sample.

The data on occupation in Table A1.3, also taken from the 1960 census, is similar to the sample distribution, except for the underrepresentation of laborers in the sample and an overrepresentation of operatives. Part of this may be due to coding errors, and part may be due to the actual change from 1960 to 1966. Otherwise, the occupational data show no upward occupational bias in the sample.

TABLE A1.3
Nonwhite Males, Northern Urbanized Areas[a]

	Census (%)	Sample (%)
Professional	6.9	7.1
Managers, Owners Proprietors	3.2	2.9
Clerical	9.1	8.3
Sales	2.0	2.2
Crafts	14.2	17.9
Operator	30.8	37.2
Service	13.8	12.2
Labor	17.6	11.5
Farm Labor	1.5	.6
Farm M, O, P	.8	0
	99.9	99.9
N	(912,000)	(1702)

[a]Source: *Occupation by Earnings and Education,* Table 2, pp. 196-219.

The Scales

The Internal Control Scale

Table A2.1 is the matrix of correlations for the internal control scale for blacks. The questionnaire contained seven items, four from Julian Rotter (numbered) and three from James S. Coleman (lettered). Two of the Rotter items were dropped because of the low intercorrelations, leaving a five-item scale. Of these five, four were administered to whites, and their intercorrelations are given in the lower half of the matrix.

The Self-Esteem Scale

The intercorrelations (Q) between the ten items of the self-esteem scale are given for both blacks and whites in Table A2.2. The ten items were used as a single scale. As an indication of the self-esteem scale's internal consistency, Kuder-Richardson Formula 20 was calculated separately for the white sample (.70) and black sample (.74). These reliability coefficients are fairly high and show that the scale is largely homogeneous. Increasing the number of scale items would increase the scale's reliability. Calculation of the Spearman-Brown Prophecy Formula, which indicates what the reliability would have been had the scale contained twenty items instead of ten, gives a coefficient of .85 for the black sample and .82 for the white sample.

There is a slight tendency for the scale to factor into two components, as shown in Table A2.3. A factor analysis (using the "minimum residuals" method to generate the factor matrices and communalities, and VARIMAX to perform orthogonal rotations on the matrices) was carried out on the white sample ($N = 655$), and two common factors were derived. The son/daughter, father/mother, and husband/wife items loaded on the first factor with loadings greater then .30; the trustworthiness, willingness to work, intelligence, and conversationalist items loaded on the second factor with loadings greater then .30. The only item that is ambiguous in its loading is "son/daughter."

The Anti-white Scale

The intercorrelations (Q) between the six anti-white items are shown in Table A2.4.

The Verbal Achievement Scale

Each item of the verbal achievement test is a 5 choice synonym test. The items form an eight-item scale; one of the items (tactility)did not scale with the others. The intercorrelations (Yule's Q) between each item and the total test minus that item are shown in Table A2.5.

TABLE A2.1
The Internal Control Scale (Yule's Q) Blacks above diagonal and Whites below

	A	B	C	1	2	3	4
A. Good luck is just as impor- tant as hard work for success.	x	.31	.33	.36	.25	.35	.13
B. Very often when I try · to get ahead, something or somebody stops me.	.21	x	.47	.29	.41	.05	.17
C. People like me don't have a very good chance to be really successful in life.	.40	.53	x	.30	.40	-.06	.09
1. Being a success is mainly a matter of hard work, and luck has little or nothing to do with it.	.35	.17	.24	x	.32	.23	.12
2. When I make plans, I am almost certain that I can make them work.					x	.18	.28
3. Most people don't realize the extent to which their lives are controlled by accidental happenings.						x	.29
4. Many times I feel that I have little influence over the things that happen to me.							x

TABLE A2.2

The Self-Esteem Scale: Associations Between Items (gamma) for Blacks (above diagonal) and Whites (below diagonal)

	Son/ daughter	Father/ mother[a]	Husband/ wife[b]	Willingness to work	Trust- worthiness	Intelligence	Conversa- tionalist	Sex Appeal	Athletic Ability	Mechanical Ability
Son/daughter	x	.54	.52	.40	.42	.56	.38	.40	.22	.15
Father/mother[a]	.58	x	.87	.48	.51	.62	.39	.28	.13	.18
Husband/wife[b]	.60	.79	x	.54	.52	.59	.38	.40	.18	.11
Willingness to work	.40	.53	.48	x	.58	.64	.32	.32	.29	.32
Trustworthiness	.61	.53	.58	.52	x	.60	.44	.36	.15	.09
Intelligence	.40	.42	.50	.38	.62	x	.68	.51	.37	.33
Conversationalist	.30	.24	.34	.26	.36	.48	x	.42	.27	.14
Sex appeal	.16	.39	.37	.15	.25	.42	.37	x	.24	.21
Athletic ability	.18	.12	.01	.17	.14	.25	.17	.21	x	.29
Mechanical ability	.04	.07	.24	.21	.21	.17	.04	.19	.27	x

[a] Asked only of respondents who are parents.

[b] Asked only of respondents who are presently married.

TABLE A2.3
Orthogonal Loadings of Self-Esteem Scale Items (for white sample, N = 655)

Item	F_1	F_2	h^2
Son/daughter	.30	.29	.17
Father/mother	.58	.20	.37
Husband/wife	.77	.12	.61
Willingness to work	.24	.38	.20
Trustworthiness	.26	.53	.35
Intelligence	.20	.48	.27
Conversationalist	.12	.41	.19
Sex appeal	.12	.29	.10
Athletics	.03	.29	.08
Mechanical ability	.06	.28	.08

TABLE A2.4
The Anti-white Items Correlation Matrix (gammas)

	1	2	3	4	5	6
How often do you feel this way when you are around whites? (frequently, sometimes, never)						
1. I am afraid I might tell him what I really think about white people. (frequently)	—	.92	.32	.49	.59	.24
2. I am afraid I might lose my temper at something he says. (frequently)		—	.59	.60	.66	.41
Agree-degree						
3. The trouble with white people is they think they are better than other people. (agree)			—	.70	.54	.56
4. If a Negro is wise, he will think twice before he trusts a white man as much as he would another Negro. (agree)				—	.66	.59
5. Sometimes I'd like to get even with white people for all they have done to the Negro. (agree)					—	.48
6. There are very few, if any white people who are unprejudiced. (agree)						—

TABLE A2.5
The Verbal Achievement Scale; Correlations of Each Test Item with Remainder of Test

Item	Q with remainder of test
Space	.62
Accustom	.54
Allusion	.48
Cloistered	.26
Edible	.69
Broaden	.69
Pact	.64
Emanate	.51
Tactility	-.16

A Note on Internal Control, Verbal Achievement, and Educational Attainment

The question of black intelligence is a much-debated, obscure issue. With the present state of the sciences of genetics, psychology, and sociology, it often seems that the best thing to do with the question is ignore it. This is what we planned to do, but unfortunately the very high association between the verbal achievement test and the internal control score makes it difficult for us to analyze internal control without at least a brief effort to clarify this relationship.

The question of which factor is cause and which is effect is usually insoluble in a survey. In this case, when we look at the verbal test score, educational attainment, and sense of internal control, it seems likely that each variable is both cause and effect of the other two. For example, bright students are more likely to finish high school and go on to college. On the other hand, it seems likely that going to school in itself raises intelligence. I.Q. is not something one is born with that remains unchanged through life.

Both high education and high intelligence should also heighten one's sense of internal control. The school teaches students to organize, to work for delayed rewards, etc. If internal control is based on recognizing the relationship between cause and effect—rejecting superstitious interpretations—then it makes sense that intelligent people should be more able to grasp this idea. The person with low intelligence may have greater difficulty imagining the future (for example, what would it be like to have money in the bank?).

At the same time, a sense of internal control should lead to higher levels of education and a higher verbal test score. The sense of internal control motivates people to save, to not quit their job, to buy a house, to avoid eviction, and it is exactly this kind of deferred gratification and future orientation which inspires students to stay in school, to acquire job skills, etc. In the same way, a sense of internal control should manifest itself in a willingness to study, a commitment to learning, a willingness to accept a teacher's direction. Over a period of years, the tested ability of the high-control student should increase.

To the extent that an increased sense of internal control does cause intelligence test scores to rise, we can argue that the racial factors that depress control also lead to depressed intelligence.

Internal control plays a different role for blacks than for whites. Figure A3.1 gives the associations (gammas) for white and black men and women. For whites, the pattern is the same for both sexes; verbal ability is closely linked to education, and the assocation with internal control is moderate. In contrast, for blacks the association between intelligence and education is much weaker, especially for women. Why is this? Our best guess is that during the 1940's and 1950's, when most of our respondents were in high school, completion of high school was not considered as useful or as necessary among blacks. Thus even successful students with high ability dropped out—to marry, have a baby, join the army, take a job, help out at home or on the farm. Thus the boys (and especially girls) who completed school might not have that much more ability than those who dropped out. In contrast, the association between internal control and verbal ability is stronger for blacks than for whites— exactly the same pattern as Coleman found for high school students. This points out, we think, the central role that internal control plays for blacks.

There is one procedure which will tell us a good deal about the process of formation of internal control and intelligence. Growing up in the South reduces both control and intelligence, and the earlier a child migrates to the North, the less the damage. We are now in a position to compare persons who migrated at different ages to see at which age the environment begins to take its toll. In the case of migration and intelligence, this will be the third time that this finding has been demonstrated, but the first time with a national sample. In the 1930's Otto Klineberg[1] demonstrated that southern black migrants who entered northern schools at kindergarten scored higher on intelligence tests than those who moved north later, after first being in southern schools. In the 1950's Everett S. Lee[2] replicated the study. Both authors were satisfied that these differences could

---

[1] Otto Klineberg (April 1963), "Negro-White Differences in Intelligence—Test Performance: A New Look at an Old Problem," *Amer. Psychologist, 18,* 198-203.

[2] Everett S. Lee (April 1951), "Negro Intelligence and Selective Migration: A Philadelphia Test of the Klineberg Hypothesis,"*Amer. Soc. Rev., 16,* 227-233.

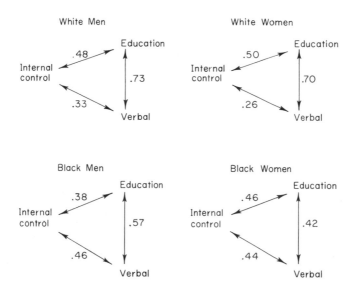

Fig. A3.1. Interrelationships of education, verbal test score, and internal control, by sex and race.

not be attributed to selective migration—i.e., any tendency for the younger migrants to be from superior families with better-educated parents.

In the same way, we can look at the effect of migration on internal control. Since this is the first time these data have been analyzed for blacks who have completed their education, we will also see the long-term impact of age of migration on the *amount* of education the migrant eventually obtains.

Let us first look at mean years of school completed, in Fig. A3.2. Men are plotted in a solid line, women with a dotted line. At the far left, the "0" age at migration refers to native Northerners. Northern-born men average almost 12 years of schooling, women about 11.6 years. The number of cases is small, so we have smoothed the curves by plotting a moving average. Thus the first point is for everyone who migrated at either 1 or 2 years of age, the second point is for everyone who moved at either age 2 or age 3, etc. (Since the number of cases here are extraordinarily small, we have presented these figures unweighted.)

Following the curves from left to right, we see that migrants who came north before starting school are approximately on a par with northern-born children. However those children who came north at age 6 or 7 finish up with much less schooling. Why is this? It seems likely that two things are happening here. First of all, these children probably did not go to kindergarten in the South, and thus are already behind their classmates. Second, they have to adjust to the trauma of moving and losing their circle of playmates and relatives at the same time that

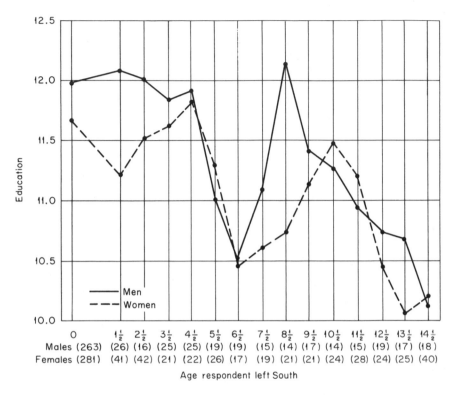

Fig. A3.2. Age of migration to the North and education, by sex. (Each point is for a two year period; unweighted N's in parentheses.)

they must make the adjustment to school. The results of this maladjustment are apparently never overcome, since this group completes on the average only 10.5 years of school.

Migrants who come north between the ages of 8 and 9 (for boys) and 10 and 12 (for girls) obtain considerably more schooling. While they are not as well-educated as those who came north as preschoolers, the differences are least for this group. This may reflect two things. First, these students were in southern schools long enough to receive some passing marks so that they entered northern schools with some self-confidence which tided them over the difficult transition. But after boys have completed 2 or 3 years in a southern school, or girls have completed the 5th or 6th grade, they are irredeemably behind. Every year they remain in the South after that further decreases their chance of finishing high school.

Only one detail of the table seems unclear. Why should southern schooling take longer to harm girls than boys? The answer may be in the fact that southern black schools (and northern segregated schools) contain a feminine bias. At least, it has always been true that southern black girls were more likely than boys to finish high school, although the sex difference has been declining recently. Thus, at the end of the 5th grade, the southern girl is much more likely to have received good grades and other praise, and thus enters the northern school with more self-confidence. It is also true, irregardless of race, that short attention span and aggressive behavior are more serious problems for young boys, and the school difficulties created by this are more serious. Hence a school environment which does not help the boys handle aggression may be more damaging in these early years.

Figure A3.3 gives the plot for the verbal test score against migration age. Again, we see that those who migrated in the preschool years perform approximately as well as native Northerners. However boys who migrate after age 5 and girls who move after age 4, have noticeably lower test scores. For girls there is

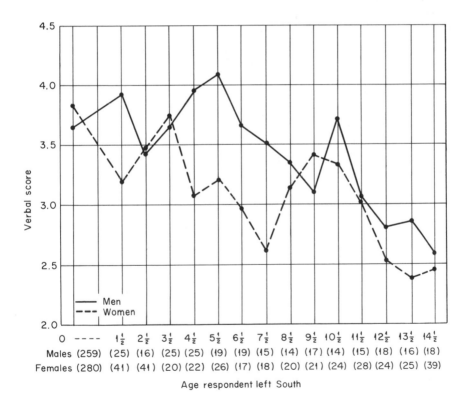

Fig. A3.3. Age of migration to the North and verbal test score, by sex. (Each point is for a two year period; unweighted *N*'s in parenthesis.)

again a low point at ages 7 and 8, followed by a resurge in the 9 to 11 age group. For boys there is a peak at age 11, suggesting that migration is easiest during latency. These reversals are not large, however, and the principal effect observable in the figures is a relentless decline in intelligence for each year of life in the South after age 4 (for girls) or age 6 (for boys).

If we are to believe the data, they suggest that the negative effects of life in the South begin even as early as the first grade. Why this should be so is not obvious—it may be the hampering effects of learning a southern dialect, the absence of radio, the repressive southern child-rearing practices, some other factor, or maybe just a sampling error. It looks as if the negative effects of the environment begin harming girls at an earlier age than boys. This could be because preschool girls are more mature than boys. With better verbal skills, they are able to relate to their environment, and be damaged by it, at an earlier age. Overall the effects of migration age is quite significant. The difference between migrating before age 4 and migrating after age 13 is approximately three-fourths of a standard deviation. (For purposes of comparison, consider that the usual white-black differences on verbal tests like this one are about one standard deviation.)

The final figure plots internal control against age moved north, and here the story is somewhat different. For men, there is no overall decline in sense of control unless the respondent remained in the South until he was 10 years old. This seems to suggest that the first effects of regional environment on internal control do not occur until the child is old enough to begin moving about on his own. And in turn this suggests that I.Q. and attitudes toward school, both of which respond to the environment at an earlier age, are the causes of control, rather than its effects.

This clarifies the picture considerably; it explains why internal control has its principal roots in parents' education and family stability. These are both factors that are important in helping the child to learn and to perform well in school. In addition, intelligence (and the child's mastering of the first major challenge outside the home—the school) is probably a major element in building a sense of internal control.

We cannot unravel this argument any further with data. There remain two possibilities. There may be some racial factor which relates to attitudes toward school and the formation of intelligence and which causes intelligence to be associated with a sense of internal control. On the other hand, it may simply be that understanding what life is about and how it works is more difficult if you are black, and hence you have to be fairly intelligent in order to cope with the system.

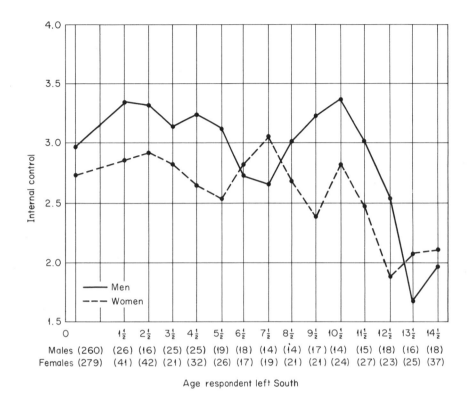

Fig. A3.4. Age of migration to the North and internal control, by sex. (Each point is for a two year period; unweighted *N*'s in parentheses.)

The Successful Men

The 21 men in the sample who could be considered very successful occupationally were selected and each questionnaire read. Below are notes on these 21 men.

The Seven Southerners

0883 is virtually an ideal case: a 32-year-old Chicagoan, an engineer working for the federal government. His background and his present situation reflect a solid pattern of achievement. Born in a border state, he was the younger of two sons; his father was a principal and his mother a teacher. He had some contact with whites and became the first black to enter his state university. He is now happily married, with two children, and rates himself very happy, with very little anti-white feeling, and a high sense of internal control, but also with high self-esteem and the ability to get angry.

The other Southerners are similar, though not quite so dramatic.

1573, a 40-year-old engineer in the Northeast, was asked if he knew where he could get a job right now if he wanted one; he casually replied "at most electronic firms." An only child from the deep South, he went to a high school named after his father; his mother also taught school. He frequently visits with whites and has managed after some difficulty to find a house in an integrated neighborhood. He did not marry until his thirties and has no children; his wife is also a professional.

1547 is a college administrator finishing his doctoral dissertation, has a more romantic and more indeterminant background. His parents had little formal schooling, but his father, a dance-band leader, was tutored at home because of childhood illness. His mother taught school, although she had only a sixth-grade education. (The respondent explained this by saying of his home town "Anything can happen down there.") He moved to a southern city at 15 and went on to Howard University. (This respondent scores "0" on the self-esteem scale, but scores at the top of the internal control scale.) He is married with three children.

1373 is the only Southerner to come from a broken home. An only child, his father died when he was 2; he lived with his mother, who taught school. He finished high school at 16 and entered engineering school in New York City immediately. He now works in ship design. He is 40, and married with one child.

191 is a 40-year-old, West Coast public school teacher. One of ten children, his father owned 160 acres in the deep South; his mother taught school. He is one of the three divorced men in the sample, and has not remarried; he has no children.

39 is a 30-year-old who grew up on his father's 200 acre farm in the deep South. After a term in the Navy, he moved to the West Coast, and after taking some college work in math, became a data processor with an aircraft company. He is discontented with his job situation, and feels that he has been a victim of discrimination. His self-esteem score is very high. He has been married three years and has one child. He has a very low internal control score.

1411 is a 26-year-old sales manager for a New York City firm and is finishing college on a part-time basis. His father was a supervisor for a southern restaurant chain, his mother a teacher. After graduating as the outstanding student of his high school, he came north to work and go to school. He is more militant than the others we have looked at so far; asked if he "is afraid he will lose his temper when talking to white people," he said yes, "all the time."

Thus we see only seven southern-educated black men in our group of 21. Of the seven, all but one are from stable homes; five had mothers who were teachers and a sixth taught part-time. Of the six fathers, three were professionals, a fourth held a supervisory post, and two owned large farms. In this sample, no one has "made it" from the southern working class to a northern middle-class occupation —at least not before the age of 45.

Two Black Immigrants

1669 is a 37-year-old engineer employed by a government agency, and was born in Haiti. His father was a lawyer. He left Haiti to go to college. Married with two children, he considers himself "very happy" but is perhaps less than satisfied with his marriage.

935 a Jamaican, owns his own electrical engineering firm. His father was a construction supervisor; his mother, a college graduate, was a full-time mother to the eleven children. He left Jamaica, first to attend college, then to work in the United States. He is the only respondent with a large family; he has six children.

Twelve Northerners

240 is another almost model case. A sales executive, he lives in a fashionable integrated suburb in the Midwest. He grew up the son of a minister (who supplemented his income with miscellaneous household work in small Kansas towns). He went to all-white schools, graduated from the state university, has a high sense of internal control, little antiwhite feelings, and calls himself a "firm believer in integration." He is married with three children.

481 is a life-time resident of a small Pennsylvania city. Like his father, he is a teacher. He attended an integrated school and had many white friends; he now lives in an almost all-white neighborhood and is strongly in favor of integration. At 29, he is one of the two respondents who has never married.

1237 was born in New Jersey and lived in a nearly all-white community. His father was a labor foreman in the city sewage-treatment plan. His mother, a high school graduate, was a matron in a local factory; she got him his first job there, and after college he returned to work in the same plant, but as an engineer. He now lives in a segregated neighborhood in a nearby large city, and seems unhappy with his life and perhaps with his marriage. He has two children, and this is his second marriage.

138 is a physician, and the second of our respondents to come from a broken home. His father, a minister, died when he was two, and his mother supported eight children by running a grocery store and selling real estate in Los Angeles. He came east to college and medical school and returned to the West Coast to set up his practice. He grew up in a changing neighborhood. Although he and his mother had many contacts with whites, he attended segregated public schools. He is married with one child.

1605 is a 37-year-old biologist with a drug firm. He grew up in a small northeastern city. His father was a theology professor, but his parents died when he was in elementary school, and he attended an all-black boarding school thereafter. He went on to college and graduate school in the Northeast.

288 is the only respondent whose parents were divorced. His stepfather, who had one year of college, was a blue-collar worker in an aircraft plant in a middle-sized midwestern city. He attended segregated elementary schools and an integrated high school. He worked for a brief period after high school, then left home to go to engineering school, and is now working in the aircraft industry. He is high in self-esteem, low in internal control, and rates himself as "not too happy."

1573 is a 34-year-old teacher in a small New Jersey city. He was born there, where his father, a poorly educated porter, died when he was 11. He lived in an integrated community (although a majority of the students in his high school were black). His mother, a practical nurse, raised five children. He rated himself only an average student in high school, but went on to obtain an education degree. He is "very happy, " low in anti-white feelings, and high in internal control.

0087 is another teacher, a native of Philadelphia. His father was a plumber and postal clerk. His parents separated when he was 8. He attended a segregated elementary school and integrated schools after that. His mother did not remarry, so apparently the family was supported by welfare. He did not marry until he was 28, and has one child.

1088 is a 22-year-old junior college graduate, who is earning a high salary as a caterer. Born in Los Angeles, his father was a janitor, his mother a social worker. He attended integrated elementary schools and his parents had a good deal of social contact with whites.

783 is a 45-year-old accountant working for the military. He was raised in a small midwestern city. One of eight children, his father worked in a steel mill. He grew up in an integrated neighborhood and attended predominantly white schools. He is married with three children. He now lives in a ghetto, is afraid to go out at night, and would prefer a predominantly white neighborhood. He has few social activities and is not too happy.

1505 is a high school graduate, 40 years old, and is well paid as a wholesale clothing salesman. He was a poor student in the nearly all-white schools in a small northeastern city; his father worked for the railroad and his mother was a domestic worker. He is also strongly pro-integration, with mild anti-white sentiments, and seems personally very secure. He married at the age of 29 and has one child.

1437 is a native of Philadelphia. His father worked for the railroad, his mother was a domestic. Although a poor student in high school, he graduated from college and is now a computer programmer. He is divorced and lives with his son. He is a militant black separatist, with a low sense of internal control, and expressed much anti-white feeling to the interviewer. One of his comments captures, in a nutshell, the problem of being militant. Asked how he would react to a white person's bad behavior toward him, he said, "Let's see now, I would kick his teeth in, black his eye, break his nose. (Anything else?) I am just kidding. I don't know what I would do."

Summary

The 21 men were selected because they had at least some college and earned over $9000 per year. In reviewing the 21, several facts emerged:

1. The occupational range is rather narrow. Only 4 of 21 earned $15,000 or more. Of the 16 college graduates, 5 are teachers, 7 are engineers, one is a computer programmer, one is a physician, one a scientist, one a business executive. Only two of these men are self-employed; the physician and one of the engineers. Four of the seven engineers work either for government or for a defense industry, and the five in education work for government. Only 3 of 21 have "line" positions in private industry—one in sales, two in management. The routes to economic success would seem to be professional training leading to a career in a field marked by tight fair-employment regulation.

2. Only 7 of the 21 were born in the South. Thus, while the average Southerner who has *not* gone to college is economically as fortunate as the native Northerner, there are few southern college graduates occupying the northern professional jobs which require college training. There is no across-the-board disadvantage to being southernborn, but there is a disadvantage at the top end of the income scale.

3. Two of the 21 were born outside the United States—one in Jamaica, the other in Haiti. Since only 2% of the total sample is foreign born, this suggests that immigrants may have a definite advantage.

4. The 12 Northerners have small-town backgrounds. None of the 12 grew up in the big ghettos of Chicago, Detroit, or New York; two are from Philadelphia. Of the remaining ten, two grew up in Los Angeles, and the remainder are from small cities—Hartford, Westchester, Pennsylvania, Union, New Jersey, Middletown, Ohio, or even smaller places. Thus only 2 of the 12 grew up in large, established northern ghettos.

Because of this, most of them attended integrated schools. Seven went to integrated elementary schools, and two others went to integrated high schools. Only three always went to "mostly Negro" schools. These percentages are typical of the whole sample, but they do not tell a complete story, for one of those three somehow grew up in a largely white neighborhood, and a second grew up in an orphanage. Only one of the three came out of segregated schools in a large ghetto (Philadelphia)—and he is the one respondent in the sample who was a "black power" enthusiast, the most anti-white of all the Northerners.

5. When we look at family background we see a fairly clear pattern. Those who do not come from integrated backgrounds come instead from high status ones. Thus, of the seven Southerners, all but one are from stable homes; five have mothers who were teachers and a sixth taught part-time. Of the six fathers, two taught, one was a professional bandleader, a fourth was a supervisor, and the other two owned large farms.

Of the three segregated Northerners, one—the black power believer—is from a poor family. The others are both from broken homes (their fathers died). One had a mother who was a real estate agent and grocery owner; the other's father had been a theology teacher (his mother also died when he was age 8, and he was raised by an uncle and in an orphanage).

Only 2 of the 19 native born successful men grew up in homes broken by separation or divorce, which is one-half the number one would expect. One of those was raised by his mother and stepfather, a relatively uncommon form of "broken" home. Three others lost a father through death, and one lost both parents—the only respondent who did not live with his mother.

Thus there are four different routes which these respondents have taken. One group are Southerners from stable, very middle-class backgrounds. A second group are foreigners, also with middle-class backgrounds. A third group are the well-integrated Northerners, from either middle- or working-class, stable or broken homes, but generally they come from smaller cities. And finally, the one working-class, big-city-ghetto product is an example of those who, lacking other advantages, can use the Muslims, Panthers, or other extremist groups as a supportive environment as they try to "beat the system."

Partial Measures of Association

The Use of Cross-Tabulation in Survey Data

With only minor exceptions, this analysis was done using the traditional techniques of cross-tabulation, rather than correlational or regression analysis. We do not want to argue strongly that cross-tabs are the right way to do analysis of social data, but we are convinced that it was the right way to do this analysis. Correlation and regression have two obvious advantages. First, they are fast procedures which generate a great deal of results quickly. Second, they handle the problems of controlling for spurious variables very well. However their very speed creates a problem. A regression equation ordinarily conceals interaction effects—a pattern where two variables correlate one way in one subgroup of the sample and a different way in another subgroup. The interaction effects can be displayed, but only after a conscious decision is made by the analyst to look for an interaction with a particular variable. But interaction effects are rare, and usually not anticipated. We should say that interaction effects are usually rare; in this study we have been swamped with them. A correlation which is positive among whites may be negative among blacks; a positive association among northern-born blacks will often be negative for Southerners; and a correlation which is positive among men will *usually* be negative among women. Internal control and self-esteem cannot be combined in an additive model, but the effect of the interaction of the two must be studied. Had we used regression, we would not have these interactions. We could have written a book in less time, and with less trouble; it might even have been a better book; but it wouldn't have been this one.

Even when cross-tabulation is used, measures of association are very valuable to reduce the data so it can be comprehended. We normally have used γ, although r, the correlation coefficient, would probably have done just as well. We do not know enough about the relationship of γ to r. Our best guess is that they are sufficiently similar that using one rather than the other makes no difference, but we are not sure of this. γ has the advantage of creating variables which are more-or-less normal. One disadvantage is that γ is sensitive to the number of categories the variable has; a γ run on a dichotomy may not look like a γ run on a "spread" variable.

The other great disadvantage of γ, the percentage difference, or the difference between two means, is that no measure of partial assocaition exists. We have invented our own, when the independent variable is dichotomous, which works as follows. Consider the example below, where age and sex are used as controls in a tabulation of high school integration and home ownership. X is the independent variable (integrated), Y the dependent (home-ownership), and Z the set of categories of one or more variables.

Z (control categories)		% owning own home (married or widowed, northern-born respondents only)	
		Integrated high school ($X = 1$)	Segregated high school ($X = 2$)
$Z = 1$: men, 20-29	$\gamma = .20$	13% (220)	9% (176)
$Z = 2$: men, 20-44	$\gamma = .30$	37% (331)	24% (105)
$Z = 3$: women, 20-20	$\gamma = .32$	11% (332)	6% (186)
$Z = 4$: women, 30-44	$\gamma = .06$	39% (375)	36% (143)

The partial is simply the weighted average of the γ's, using the geometric mean of the N's in that row as a weight. Thus the formula is:

$$\gamma_{XY \cdot Z} = \frac{\displaystyle\sum_{Z=1}^{Z=n} \gamma_{XY \cdot Z=i} \sqrt{(N_{X=1, \, Z=i})(N_{X=2, \, Z=i})}}{\displaystyle\sum_{Z=1}^{Z=n} \sqrt{(N_{X=1, \, Z=i})(N_{X=2, \, Z=i})}}$$

If, for example we wanted the partial γ between integration and ownership, for men only, using age as the control, the arithmetic would be:

$$\gamma_{XY \cdot Z} = \frac{[(.20)\sqrt{(220)(176)} + (.30)\sqrt{(331)(105)}\,]}{\sqrt{(220)(176)} + \sqrt{(220)(176)}} = \frac{(.20)(196.7) + (.30)(186.4)}{(196.7) + (186.4)}$$

$$\gamma_{XY \cdot Z} = \frac{91.27}{383.1} = .24$$

The same formula can be used with any number of control variables, each spread into any number of categories. For example, if age were three categories, there would be three terms in the equation above. Or, if we wanted a single partial γ to summarize the relationship for both sexes, we would use all four categories of Z to give us four terms in the numerator and denominator of the equation.

The same formula can be used with any measure of association, as long as the independent variable is dichotomous. In Chapter 9 a partial for the differences in two means was computed, and in Chapter 11 a partial percentage difference is used.

Index